The Oak Ridge Story

"OAK RIDGE WILL HAVE A UNIQUE PLACE IN HISTORY. IT WILL BE A LANDMARK IN THE FIELD OF ATOMIC DEVELOPMENT."

GEN. L. R. GROVES, U. S. A. (Ret.)
Commanding General
Manhattan Project, 1942–47

THE

OAK RIDGE STORY

The Saga of a People Who Share in History

By

GEORGE O. ROBINSON, JR.

Illustrated

SOUTHERN PUBLISHERS, INC.

KINGSPORT, TENNESSEE

Dedicated

. . . to the thousands upon thousands of scientists, engineers, construction men, operations managers, military personnel, workers in the plants, laborers and administrative personnel, whose skill, knowledge, dogged determination, teamwork and silent perseverance made possible the unprecedented and dramatic accomplishment which brought a new era to the world . . .

. . . to the thousands who continue their labors to maintain the leadership of the United States in the new field of atomic energy;

and to

BILLIE

WITHOUT WHOSE ENCOURAGEMENT THIS BOOK
WOULD NOT HAVE BEEN WRITTEN.

Introduction

IN World War II, when the Japanese were in control of
Southern Luzon, they conducted a roundup of persons
suspected of unfriendly attitudes. One of these was an elderly
American who had resided in the Philippines for many, many
years. A Japanese officer questioned him about his nation-
ality. The American replied that he was from Tennessee. A
perplexed look crossed the officer's face. Then he decreed:
"You may depart. You are of a non-belligerent nation. Japan
has no war with Tennessee."

On August 6 and August 9, 1945, the Japanese realized in
a most dramatic manner that they did indeed have a war with
Tennessee, as well as the rest of the United States. On those
dates, two atomic bombs shattered the cities of Hiroshima
and Nagasaki. One was supplied with a new kind of explosive
material produced in unprecedented facilities at Oak Ridge,
Tennessee; the other with a new kind of explosive material
produced at Hanford, Washington. The Atomic Age had
broken over an astounded world.

For sheer, stark drama, no single venture in the Twentieth
Century approaches the development of atomic energy—first
as a weapon for use in World War II and, in subsequent
years, as a tool to benefit mankind. Americans should not
only take deep and grateful pride in the accomplishment but

should acquaint themselves with the general history of the development and its promises of a brighter future. The good to be derived by humanity from atomic energy will, in time, completely eclipse the destructive character of one of the greatest advances made in all the long history of man's search for knowledge.

During World War II, three great main installations were established by the United States Government to bring about in the shortest possible time the development of a new weapon to be known as an atomic bomb. These were at Oak Ridge, in Tennessee, Los Alamos, in New Mexico, and Hanford, in the State of Washington. Smaller development areas were established elsewhere. The Nation's great industrial organizations and universities, aided by thousands of other organizations and groups of varied description, led the way in breaking through many barriers to reach the new continent of atomic energy. The successful accomplishment of the atomic energy mission is attributed to what is probably the most comprehensive cooperative effort in history on the part of Government, science, business and labor, and individuals in all walks of life working in a common cause on a single project.

Of the three major installations, Oak Ridge, Los Alamos and Hanford, which were built from scratch, including large communities to house atomic energy workers, Oak Ridge has a particularly fascinating and unique place in the modern-day story of the atom; the Oak Ridge Area, which rose from what was once a sparsely-settled section in the foothills of the Cumberland Mountains in East Tennessee, has in truth come to be known as the Symbol of the Atomic Age.

Interestingly enough, atomic energy is not as rare or as strange as generally thought. It is not a thing of "black magic."

[8]

Atomic energy is the basis of life itself. It is with us every minute, every hour. Consider the sun, in which a nuclear, or atomic, reaction has been going on for millions of years, where hydrogen atoms convert (fuse) into helium under high temperatures (at least 1,000,000 degrees Fahrenheit) with simultaneous release of enormous amounts of energy. Energy from the sun nurtures and sustains man and all other life on the earth. The sun is the greatest of atomic energy "factories."

But we leave to others the technical explanations of atomic energy. "The Oak Ridge Story" gives attention to the historical phase of Oak Ridge itself. It is the story of the birth pains of a city and a project, of a people who share in history, the human side of the startling atomic adventure, an accounting of Americans at their best when the chips were down. It is the story of the pioneers of another era giving up their lands and homes so that a new type of pioneer coming from every corner of America could bring about a New Age and how those who participated in the ushering in of this New Age lived and worked as great plants and a city arose almost miraculously from once pine- and oak-dotted terrain.

And it is the story of how these new pioneers, under the general supervision of the United States Atomic Energy Commission, are now striving to bring to rapid fruition the great peacetime potentialities of atomic energy. Oak Ridge, sometimes called the "Cradle of the Atomic Age," is at once the Crossroads of a New World.

In bringing into one complete work the story of Oak Ridge and its people, the author wishes to make grateful acknowledgment to Mr. Clifford Seeber, of Knoxville, Tennessee, a native of the Oak Ridge Area, for the use of certain historical facts contained in a paper entitled "From Acorns to Atoms." The author also has utilized background information contained in articles appearing in the semi-official project newspaper, the *Oak Ridge Journal,* which served the community from 1943 until the Spring of 1948.

Releasable facts on Oak Ridge plants have been carried in official releases of the War Department's Manhattan Project, the United States Atomic Energy Commission and various contractors of the Manhattan Project and the Atomic Energy Commission, and in various other public accounts. In addition, the author has been privileged to occupy a ringside seat for the unfolding of "The Oak Ridge Story," becoming associated with the atomic project in June 1943, when it began to burgeon forth in all its splendor, and has drawn from his own knowledge and observations of the development.

GEORGE O. ROBINSON, JR.
Oak Ridge, Tennessee
September 27, 1950

Contents

[15]

Illustrations

[13]

The Oak Ridge Story

CHAPTER I

The Prophet

NEAR the city limits of Oak Ridge, Tennessee, not far from a high wire fence running forbiddingly along one of the main traffic arteries leading into the community, a lone grave rests on a grassy, overgrown knoll. It is marked by an inconsequential little stone—scratched and scarred.

Resting beneath this forlorn marker in the shadow of a city of 30,000 where live the personnel who grapple daily with the new force of atomic energy are the remains of one John Hendrix. Many years ago, John Hendrix, an ascetic, roamed the fields and woods of a sparsely-settled, isolated section of East Tennessee. John Hendrix had visions. He recited these visions frequently. Those who paused to listen often laughed.

John Hendrix assumed the role of a prophet around the turn of the Twentieth Century, when about 50 years of age. His visions manifested themselves more acutely as he would return from a communion in the deep woods which stretched along the ridges near his stark, weather-beaten home. One Spring day, shortly after returning from a meditative walk, he startled his wife with the prediction that in a few years a railroad would be built from Knoxville, 20 miles to the east, through the central part of Anderson County, their home.

The buxom woman looked at him speculatively and went

about her business of housekeeping. Her indifference did not upset him. On the contrary, he became more bold and open with his prognostications. Well, he might, for the Louisville and Nashville Railroad within a few years began construction of a new line along the approximate route spelled out by Hendrix.

Hendrix began to enlarge his horizons. He found his most ready audiences at the crossroads' store near his home and to them he solemnly voiced his prophecies. There was no historian to put his stories to paper but fragments of information remain which give accountings of John's speculations of the future.

Of two small farming communities near his home, Scarboro and Robertsville, Hendrix foresaw unusual things. On one occasion, he emerged from his beloved woods with a strange and wondrous story.

"In the woods," he said, "as I lay on the ground and looked up into the sky, there came to me a voice as loud and as sharp as thunder. The voice told me to sleep with my head on the ground for forty nights and I would be shown visions of what the future holds for this land."

Hendrix did what the "voice" told him to do. He wandered through the thickets and the briers until he came to a clump of trees which satisfied him as a spot for his tryst. He put his head on the ground and slept a fitful sleep for forty nights. The rains came but he was unflinching. On the forty-first day, he came forth to tell of the vistas which had been opened up to him. Part of the time, he said, had been spent in Paradise.

". . . and I tell you," Hendrix said to his neighbors gathered at the crossroads' store, "Bear Creek Valley some day will be filled with great buildings and factories and they will help toward winning the greatest war that ever will be.

[18]

The slab over Hendrix' simple grave near the town of Oak Ridge

"And there will be a city on Black Oak Ridge. The center of authority will be on a spot which is middleway between. Sevier Tadlock's farm and Joe Pyatt's place.

"A railroad spur will branch off the main L & N line and run down toward Robertsville and then it will branch off and turn toward Scarboro. It will serve the great city I saw in my vision.

"Big engines will dig big ditches and thousands of people will be running to and fro. They will be building things and there will be great noise and confusion and the earth will shake. . . ."

"I've seen it, it's coming," he would mutter and then stalk away to the summit of Pine Ridge, where he would gaze across the valley toward Robertsville and meditate and pray again.

John Hendrix died in 1903, when Theodore Roosevelt was President and the Wright Brothers made history with the first successful mechanical airplane flight from Kill Devil Hill on North Carolina's seacoast, four miles south of Kitty Hawk. They buried the prophet in a clearing near his home. Forty years later, there arose a short distance away the city he predicted would some day rest along one of the ridges—Black Oak—over which he roamed and dreamed.

In the county of Anderson in the state of Tennessee, seven miles from the Anderson county-seat town of Clinton, a bustling community of 30,000 goes about its business today along Black Oak Ridge. In every corner of the world, Oak Ridge is the symbol of a new era, the Atomic Age.

The old communities of Scarboro and Robertsville, through which John Hendrix walked his weary way, and a third farming settlement, Wheat, several miles to the southwest, have been enveloped. A railroad spur runs into the city from the main line of the Louisville and Nashville Railroad.

A huge Administration Building, the center of authority for the new city, stands where Hendrix said it would.

In Bear Creek Valley, there is a great and imposing plant known as the electromagnetic facility, built to produce uranium-235. Here was produced for the first time in history in a significant amount the material which went into the world's first atomic bomb utilized in warfare, the devastating weapon which cascaded over Hiroshima, Japan, at 8:15 A.M. on August 6, 1945, taking 80,000 lives. Three days later, another atomic missile fell on Nagasaki, Japan, taking 73,000 lives. The Japanese had had enough. Surrender followed.

John Hendrix would not recognize his land today if he were to return. He sleeps on. But his spirit must have stirred uneasily early on the morning of July 16, 1945, when in the desert stretches of Alamogordo, New Mexico, a huge blast shook the earth and the sky became lit up as if by the noonday sun. The test of the world's first atomic bomb had been successful, an event described by Brig. Gen. Thomas F. Farrell, an observer, as follows:

"The whole country was lighted by a searing light . . . golden, purple, violet, gray and blue. It lighted every peak, crevasse, and ridge of the nearby mountain range with a clarity and a beauty that must be seen . . . to be imagined. It was that beauty the great poets dream about but describe most poorly and inadequately. Thirty seconds after, the explosion came first, the air blast pressing hard against the people and things, to be followed almost immediately by a sustained, awesome roar which warned of Doomsday and made us feel that we puny things were blasphemous to dare tamper with the forces heretofore reserved to the Almighty."

It was a solemn moment in world history. The roar reverberated in time back along Bear Creek Valley in the hills of East Tennessee. . . .

CHAPTER II

Strangers in the Valley

A PARTIALLY cloudy sky greeted East Tennessee on September 19, 1942. As the day wore on, a gentle southeast
wind helped to alleviate the 91-degree temperature. In Eu-
rope, the Russians were striking vigorously at the Nazis on
the Volga in their defense of Stalingrad, and the late Wendell
Willkie was a visitor in Russia as a guest of the Soviet Government. In the Pacific, General MacArthur's heavy bombers
and fighter planes were harassing key Japanese bases on New
Britain Island and in Northern New Guinea. Announcement
was to be made soon that Admiral William F. Halsey had
been placed in charge of the Navy's Task Forces in the South
Pacific.

An entire nation bit its collective lower lip in grimness.

Throughout the United States, the loins of war were being
girded at a faster pace. The know-how that is uniquely American was beginning to assert itself in many places throughout
the land. Scrap drives began to get underway. The pinch of rationing was being felt more widely. The tramp of thousands
of feet began to be heard with more regularity in training
camps from the Atlantic to the Pacific and Army maneu-
vers were getting underway in Middle Tennessee. Simultaneously, Lt. Gen. Brehon Somervell of the War Department's
General Staff was telling Americans that "we are not going to

[21]

win until everybody puts everything he has into the war."

Life in the farm homes, the schools, the churches and the rural stores along the ridges and in the valleys which make up Anderson County and the adjoining County of Roane went on about as usual. Only the departure of men to volunteer, to answer the call of their draft boards or to leave for employment at the great sprawling plant of the Aluminum Company of America at Alcoa, 34 miles away, and at other industrial plants in East Tennessee, disturbed the normal life of the section.

None knew—actually only a chosen few in Government circles in Washington did know—of the establishment 37 days before, on August 13, by order of the Commander-in-Chief, Franklin Delano Roosevelt, of a mysterious segment of the War Department's Corps of Engineers designated simply "The Manhattan District."

Since August, however, residents of Anderson and Roane counties had noticed a few strangers roaming the valleys, the ridges and the gaps which made up their homeland. The newcomers were sighting through instruments, surveying and measuring. They were engineers, dressed in rough khaki, wearing high boots, taciturn and noncommittal. Asked what they were running lines for, they would answer. "For seventy-five cents an hour." The natives wondered and talked about it among themselves. They made wry attempts to shrug it off. "Only more Tennessee Valley Authority people," they would say.

But just as noonday came on September 19, five men, two of them high ranking officers in the United States Army, with the Corps of Engineers' insignia on their blouses, stood at a vantage point at the railroad whistle stop of Elza in Anderson County. . . .

The spot from which they looked out over a long valley

had seen the tides of an empire swirl about it. It had echoed
to the rumblings of the covered wagons from the East and the
staccato sound of rifle fire as the Indians retreated before the
white men. Long before, a race whose identity is buried in
antiquity's archives had built mound cities in the blue grass
basins to the south and southwest.

The State to which the five had come to make a momen-
tous decision was the seed-bed of the civilization of the Old
Southwest; of the development of that West, a historian has
said: "Its center rested in Tennessee, the region from which
so large a portion of the Mississippi Valley was settled by descendants of the men of the Upper South."

In Tennessee, history walks with a heavy foot.

To the northeast of the point where the five visitors stood
on September 19, 1942, is Greeneville, resting place of the
country's 17th president, Andrew Johnson, the tailor, who
succeeded to the office upon Lincoln's death; to the southwest is Columbia, home of James K. Polk, the 11th president;
to the west near Nashville is The Hermitage, where Andrew
Jackson, "Old Hickory," the 7th president, died in 1845; to
the south is Chattanooga, near where the famous Civil War
battles of Chickamauga, Lookout Mountain and Missionary
Ridge were fought in 1863; to the southeast is Sevierville,
named for John Sevier, Tennessee's first governor; far to the
southwest is Memphis, where the Spaniard, DeSoto, is said by
some historians to have first seen the Mississippi River in
1541; to the northwest is Jamestown, home of Sergeant Alvin
York, World War I hero, and close by, a short distance from
Byrdstown, is the birthplace of Cordell Hull, former United
States Secretary of State, who served longer in that position
(March 3, 1933—November 21, 1944) than any other man in
the nation's history.

Illustrious names shine in the history of Tennessee, the six-

[23]

teenth state admitted to the Union (June 1, 1796) and first settled by a white man in 1769, when William Bean built a log cabin near the junction of Watauga River and Boone's Creek in East Tennessee. These names include Sam Houston, "The Raven," one-time governor of Tennessee who later became the Liberator of Texas; David Crockett, pioneer backwoodsman and later a Congressman; David Glasgow Farragut, the first Admiral of the American Navy, who was born 11 miles southwest of Knoxville; Nathan Bedford Forrest, the Confederate general whose battle strategy was "Git there fustest with the mostest men"; William C. Claiborne, whose vote made Thomas Jefferson president over Aaron Burr, and E. E. Barnard, world-famous astronomer.

Nature, too, had erected monuments of its own in Tennessee. To the east and southeast of the spot where the men stood are the Great Smokies, the country's greatest mountain mass east of the Black Hills of South Dakota and one of the oldest mountain ranges in the world, where mountain men still scrape their fiddles in lively old ballads handed down from generation to generation, and pioneer crafts of spinning, weaving and basket making still survive; to the west is Reelfoot Lake, created in 1811 in the spasms and convulsions of an earthquake when the bottom of the Mississippi River seemed to fall through as it flowed backward into a terrible abyss of thousands of acres; and to the south is Lookout Mountain, a rock-faced promontory carved by the currents of the Tennessee River and overlooking Moccasin Bend at Chattanooga. . . .

. . . Their conference over, the five visitors to the valley entered their automobile. Their destination was Knoxville, 20 miles to the east, where a metal slab on a building at the corner of Gay and Church Streets carries this legend: "SITE OF THE BIRTHPLACE OF TENNESSEE—HERE THE FIRST CONSTITU-

Famous Norris Dam of the TVA helps furnish power for Oak Ridge

Tennessee Conservation Department

The Great Smoky Mountains (Gatlinburg in the foreground) are near Oak Ridge

TIONAL CONVENTION WAS HELD IN THE OFFICE OF COL. DAVID
HENLEY, AGENT OF THE WAR DEPARTMENT, KNOXVILLE—JANU-
ARY 11–FEBRUARY 6, 1796." Upon reaching the city, they went
directly to their hotel. One of the officers lifted a telephone
in his room, asked for long distance. When he reached his
party in the War Department in Washington, he said:

"The location tentatively selected in July and which has
been under survey is ideal for our purposes. We should proceed with land acquisition."

The other colonel and the three companions—they were
officials of the Stone and Webster Engineering Corporation
of Boston, under contract with the War Department to carry
out in high secrecy the development engineering, design engineering and construction of certain plants for producing
uranium-235—nodded in agreement. Their faces were grim,
their demeanor serious; another step had been taken into a
dark, foreboding and uncharted path.

The site they had chosen for the building of fantastically
big facilities which had no pilot plants to serve as models, and
which at that time existed only in laboratory conceptions and
in blueprints being drawn under the supervision of America's
most distinguished scientists—the proposed plants had no
counterparts anywhere in the world—was a tract of 92 square
miles, 58,800 acres on the Clinch River in Anderson and
Roane counties 18 miles west and northwest of Knoxville.

(The Oak Ridge Area presently constitutes 58,762 acres of
which 30,315 are in Anderson and 28,447 in Roane. The area
is approximately 17 miles long, averages 7 miles in width and
runs from northeast to southwest. While additional acreage
was purchased after 1942, other land has been sold or relinquished. Of Anderson County's 342 square miles, the Oak
Ridge Area occupies about one-sixth.)

In 1933, another revolutionary step in the American way

of life had taken place, the authorization by the Congress of the Tennessee Valley Authority. The first of its great dams, Norris, and the city of Norris, which were built in the middle 1930's, stood 16 miles northeast of the new War Department site. One of the contributing factors to the decision made September 19 was the TVA, for great amounts of power were needed for what the Manhattan District proposed to do. In addition to sufficient electric power, Clinton Engineer Works, as it became known, had to be safe from air attack, not too close to large centers of population, large enough to accommodate four separate plants with flat building areas separated by natural barriers, accessible to rail and motor transport, and on land of reasonable value adjacent to a dependable water source. Specifications also called for space for a town, a local labor supply and topography which would facilitate security control.

For the second time within a decade, the wheel of Fortune spun toward Anderson County, Tennessee. Anderson County's first big drama was the TVA; the second, the Clinton Engineer Works, was so vast in scope that it included part of Anderson's neighboring county of Roane, where the first coal was mined in Tennessee in 1814.

Shortly after the dramatic telephone call to Washington, the deluge fell upon the 1,000 families in the area selected. They were to lose their homes and their farms. Their land, they were told, was needed immediately for the war effort. They were further informed that efforts would be made to relocate them satisfactorily, that "adequate compensation" would be made for their land, property and crops and that "every consideration would be given the people by their Government." Among such considerations was the taking over of the care of 65 cemeteries on the project site. The plots are still looked after by the Government. By November 1, 1942, all property owners had received notices of condemnation, ordering them to vacate the project site not later than January 1, 1943.

The race into the unknown was pressing on, men were walking an untrod path along the brink of disaster; time could not stand still. In Washington, the question of what direction Germany might be moving in scientific fields furrowed the brows of key officials.

Concurrently with the initial trickle of official personnel onto the project site in October, Knoxville newspapers and the weekly newspapers at Clinton, in Anderson County, and Harriman, in Roane County, began to ask for details of the project, its purposes and objectives. After much probing and questioning locally and in Washington, local headlines announced:

"Army surveys for project in area. Telegram from Congressman Albert Gore announces that about 56,000 acres of land in Anderson and Roane counties will be used as site for a demolition range." A demolition range, it was explained, is an area in which targets are set up to be destroyed by artillery fire or airplane bombings. Indeed, for several weeks, Clinton Engineer Works was known as the "Kingston Demolition Range"—a prophetic designation.

Kingston, county seat of Roane County, was the western outpost of the United States Army shortly before the American Revolution and famous as the site of a huge breastwork built by the Indians in prehistoric times. It had a brief fling in the nation's history on September 21, 1807, when Tennessee's General Assembly named it as the State's capital for one day, only to move the capital back to Knoxville the next. Again in 1843, Kingston was named by the Tennessee Senate as the State's permanent capital but the House of Representatives chose Murfreesboro. This necessitated a compromise —Nashville, the present seat of state government.

The site having been chosen for the Clinton Engineer Works, the once tranquil, isolated valleys of Anderson and Roane counties changed overnight. While Stone & Webster engineers directed the cutting of ditches, the driving of stakes in barnlots and the opening of roads across cornfields, and as huge trucks, the first of thousands to come, bumped across open terrain, the first official headquarters office for Clinton Engineer Works was opened in a room in the Andrew Johnson Hotel in Knoxville at 8 A.M. on October 26, 1942.

First Government man on the job for keeps was Robert J. Dunbar of Osceola, Iowa, an engineer with the Corps of Engineers, who arrived at the temporary offices 15 minutes before Lt. Col. Warren George, who served as first construction engineer for the project. Major Thomas J. Rentenbach of Hancock, Michigan, was the second Army representative to report. First Stone & Webster official to report was Talley W. Piper, personnel manager, October 26, 1942; five days later he was joined by a brother, J. P. Piper, procurement officer. In a few days, the first shovel of dirt in project construction was turned by the firm of Walters and Prater, of Morristown, Tennessee, in the building of a railroad siding at Elza.

Responsibility was placed on the Ohio River Division of the Army's Corps of Engineers for acquiring the land, which eventually cost the Government approximately $2,500,000. Dissatisfied with the first appraisals, most of the landowners carried appeals to the Federal Court in Knoxville, where juries generally increased the Government's offers. In the final settlements, average cost of an acre of land for the Oak Ridge Area was approximately $45.

Preparation for construction and actual construction work

could not, however, await determination of the land suits, some of which were not settled until 1944. On November 2, 1942, the first carload of materials arrived. From then on until the Summer of 1945, wondrous things happened.

The manner in which the Government acquired the land for its venture had reverberations in Washington. In the Summer of 1943, a House Military Affairs Sub-committee headed by Representative Clifford Davis of Memphis investigated complaints and reported that "fairer prices should be paid those forced from the area by the Government." Former Representative John Jennings, Jr., of Knoxville, who requested the probe, told his constituency that "the Secretary of War has assumed the guise of an invader," a charge which he undoubtedly would have softened if Henry L. Stimson had been officially privileged to discuss the matter with him.

But there was at least one person among the 1,000 families —3,000 persons—forced to depart from the area who felt that Mr. Jennings knew what he was talking about when he mentioned "invader." He was a part time manufacturer and dispenser of the potent white beverage known as "moonshine." For some years, it seemed, the Government had been singling him out for especial attention.

Born and reared in the Great Smokies, where he was living happily on Roaring Branch and plying his trade with great finesse, he found himself evicted in 1928 when the United States Park Service condemned and bought up a huge section of land, including his 15 acres, for a National Park. Casting about for a good location for business, he selected Union County. But he hadn't calculated on the Tennessee Valley Authority, which condemned and acquired his land in the middle 1930's to help hold back the waters of Norris Dam.

His next stop was Anderson County. But he landed in the wrong place again. The Government needed his land for the

[29]

Clinton Engineer Works. It is not recorded if he found a new business site. But he did remark philosophically as the third condemnation notice in a period of 14 years arrived:

"If I knew a place where there warn't no Government men but only revenoors to dodge, I'd shore go there."

For others, a leather-faced oldtimer put the problem of eviction into more understandable language. "The only difference," he said petulantly, "is when the Yankees came before, we could shoot at them."

—and That's Not All

In addition to being the site of the giant production facilities which produced the first U-235 in world history for use in an atomic weapon, Tennessee has recorded other notable "firsts."

In 1801, Tennessee passed the first state law against dueling. The death penalty was provided.

In 1819, the first periodical against slavery appeared at Jonesboro, being succeeded by the Emancipator in April, 1820.

In 1831, the first publication in the United States devoted principally to railroading, "The Rail-Road Advocate," appeared at Rogersville.

The first town in the United States to be named in honor of the maiden name of George Washington's wife, Martha Dandridge, is Dandridge, in Jefferson County.

In 1866, the Ku Klux Klan was first organized at Pulaski.

In 1925, Dayton was the scene of the famed "Monkey Trial" when William Jennings Bryan and Clarence Darrow played a great drama while the world took sides on the theory of evolution and debated whether John T. Scopes, a high school instructor, was in the right when he violated the Tennessee law prohibiting the teaching of the theory. Scopes was fined $100; the law's constitutionality was upheld by the Tennessee Supreme Court on appeal and remains on the statute books.

CHAPTER III

Of Those Who Left

WHAT is now the Oak Ridge Area was settled in the late
1700's and the early 1800's by hardy and rugged individuals who helped push back the frontiers of America.

The majority were members of that vast English, Scotch
and Irish clan which first followed the Quakers into Pennsylvania, then pushed down into Virginia and North Carolina.
Later, they followed the mountain gaps—Cumberland Gap
on the Tennessee-Virginia line was one of the first passage-
ways—into the new, promising, free land in the West. With
them came the Dutch and the Swedes, a melting pot made
up of resourceful, peace-loving people, pioneers in frontier
democracy.

Land grants given by the Government also attracted some
Revolutionary soldiers, among them some who had fought in
the famous battle at King's Mountain, on the North Carolina–South Carolina border against the British. One Revolutionary soldier who settled in the Oak Ridge area was Captain John Harrell, who had accompanied Washington when
he crossed the Delaware. His descendants today are scattered
over a broad East Tennessee section.

In addition, a scattering of German immigrants appeared
in the area around the turn of the 19th Century. Probably

Clinch River meanders as miles around the atomic project

the most famous of these was Frederick Sadler, a wagon maker, who arrived in Pennsylvania in the late 1700's, reared a large family of girls and then decided to move westward. In covered wagons, he made the trek with his wife, eight daughters and eight sons-in-law and settled in a little valley near the Oak Ridge section. The family names of seven of his sons-in-law (the name of the eighth was not preserved)—Leinart, Spessard, Leib, Shinlever, Claxton, Clodfelter, and Bumgartner—mingle with the more common names of Smith, Jones, Brown, Watson, Gallaher, Jett, Standifer, Cross, Wilson, Browder, Morton, Stooksbury, Haun, Brennan, Price, Harmon, Dunlap, McKinnon and Reed in Tennessee today.

When the War Department put its finger on the now famous acreage in Anderson and Roane counties, the contrast in the way of living was as sharp as that found in any section where there are green valleys and rocky ridges. In the valleys were substantial homes on farms ranging from 200 to 600 acres, rolling fields, hard roads, electric power and modern equipment. On the ridges, the foothills of the Cumberland Mountains which rise to the West, were eroded hills, tenant houses, sedge grass pastures, scrub stock, hound dogs and cabins perched on rock formations. But in the entire 58,800 acres there were only two one-teacher schools, serving about 30 children each. In the communities of Scarboro, Robertsville and Wheat, however, there were consolidated schools accommodating around 1,100 pupils.

Along the perimeter of the area selected stood other settlements known as Elza—named for another German family which pioneered in East Tennessee—Solway, Blair and Edgemoor.

Elza, Edgemoor, Blair, Solway and Oliver Springs, named for a community six miles west of the area, were to become especially familiar designations in the Atomic Age. Through the barriers which were established at these points, and at two others—Gallaher and White Wing Entrances—to control visitors into the town and the vast plant areas, world-renowned scientists and the great men in American military, business, industrial and Government circles passed during the hectic days of construction and first plant operations. And for month upon month during World War II, many American manufacturers, suppliers and jobbers scratched their heads in bewilderment, for orders for materials from the War Department simply specified shipments to various engineering and construction companies at Elza, Tennessee and Blair, Tennessee. The bewilderment was understandable; the Postal Guide in 1942 listed neither settlement.

William Tunnell and his family were the first settlers in what is now Oak Ridge, arriving from Virginia around 1792. They occupied land where Robertsville later stood. The Tunnell family and that of Anne Howard, an English pioneer, was joined by Isaac Freels, an Irishman of Presbyterian faith, who entered 1000 acres of land adjoining the two. Then came the Peaks, the Lees and the Garners. A member of the Garner clan was former Vice-President John Nance Garner's father, who migrated to Texas after the Civil War.

Two other illustrious men in American life had their roots in the immediate vicinity of Oak Ridge. Sam Rayburn, speaker of the National House of Representatives, lived as a young boy across from Clinch River west of Wheat in Roane County before his family moved to Texas. Just before the late William Gibbs McAdoo, U. S. Secretary of the Treasury (1913–19), was born in Georgia during the Civil War, his father had lived in Clinton. After the war, the McAdoos returned, and Clinton technically claims him today as its native son.

In the 1840's, the Gallahers came in from the East to buy

up 1500 acres of river bottom and gently rolling hill land along Clinch River in the west portion of the Oak Ridge area. The Gallahers became one of the outstanding families in East Tennessee and through the years have supplied bishops, judges, ministers, politicians and business men throughout the United States. The Gallahers can trace their family tree back to the moors of North Ireland and the House of O'Gallcobair and then back into the Seventh Century when old King Callack ruled Ireland.

The name of Gallaher is synonymous with the community of Wheat, near where there arose during World War II one of the largest industrial buildings in the world—the gaseous diffusion plant for the production of uranium-235. This building covers 44 acres and is nearly half a mile long.

On February 9, 1947, the late William Gallaher, one of the early residents whose land was taken but who remained with others to work on the "project," told a nation-wide radio audience on the "We, the People" show emanating from Oak Ridge:

"I was born in the house my grandfather built back in 1846 (when the United States and Mexico began war). This is mighty pretty country around here—the Great Smoky Mountains to the east and the Cumberland Mountains to the west. Don't blame my grandfather a bit for settling here.

"All the folks in these parts were farmers. They worked the ground and minded their own business, peaceful folks living a simple life. Of course, when the Civil War came along, we sent a few of our boys out to fight. And then in World War I we did our share. But other than that, we didn't pay much attention to the outside world and they didn't bother with us. That was up to 1942, anyway, when one day a man came to our house and said he was from the Government. 'We're going to buy up your land,' he said to me. 'All of it?' I asked.

'Yes, sir,' he said, 'we're going to buy all the land in this section. Everyone has to go.'

"I went outside the house with the visitor and looked around me . . . up at the green hills my grandfather had come across 100 years earlier, and I looked at the farm I'd worked for half a century. I asked the visitor what the Government was going to do and he said he didn't rightfully know, but it was for winning the war. I had three sons in the Service —two overseas—and I figured if giving up my home and my land would help. bring them home sooner, I'd be happy to do it. . . ."

The hamlet of Wheat, named for its first postmaster, Frank Wheat, was originally known as Bald Hill. It has furnished illustrious leaders in many walks of life in Tennessee and elsewhere. The names of Cross, Cox, Brittain, Jones, Christenberry, Rigsby, Arnold, Waller, Green, Hembree, Driskell, McKinney, Young and Davidson and many others figure prominently in the life and times of Tennessee.

A subscription, or "loud" school, so-called because of the custom of reciting lessons aloud and in unison, was established in the Wheat Community in 1876 by the Rev. John P. Dickey, a Methodist minister. The school was held in an old log building and marked the beginning of a movement to provide educational facilities for the section.

Two years after the founding of the "loud" school, Poplar Creek Seminary was founded and headed by the Rev. W. H. Crawford, Cumberland Presbyterian minister. Some time later, Dr. C. W. Butler, Presbyterian pastor, who also was a skilled physician and a Princeton graduate, was associated with Mr. Crawford as a teacher in the Seminary. Courses were similar to those offered by state high schools in later years.

In 1879, George Jones, a man of generous spirit, presented the Seminary 200 acres of land encircling a four-acre plot owned by the local Baptist Church. The deed contained the unique stipulation that the Seminary Board of Trustees allow all persons in whose homes a student or students stayed to build on and enjoy full use of one-acre lots. If students no longer lived in the home or if the householder committed an act detrimental to the school, the resident was obliged to sell his house to a qualified person. Many persons were attracted to the Seminary through this building policy.

In 1886, Poplar Creek Seminary was chartered by the State of Tennessee as Roane College. Students were attracted from over Tennessee and neighboring states to the four-year liberal arts college which offered both Bachelor's and Master's degrees. Roane College virtually ended, however, in 1908 when Wheat High School was established, but the Board of Trustees continued in control of the property until 1916 when it was transferred to the Roane County Board of Education. After 34 years, the old Wheat High School, which replaced Roane College, was demolished in 1950.

What was once Wheat and the site of a college is today in the restricted portion of the Oak Ridge Area. Visitors generally are not allowed within these confines. There is, however, one exception. In the shadow of the massive gaseous diffusion plant, a reunion is held each year at the George Jones Memorial Church, which adjoins a cemetery in which are buried some of the early settlers. The annual Homecoming of former Wheat residents is probably the most unusual gathering held in the United States. Those attending are admitted through the barriers to the restricted area of the atomic energy plants only through passes approved by the United States Government. And near the scene of the festivities, armed guards patrol the roads and an eerie calm hangs in the atmosphere.

Clinton, county seat of Anderson County, from which

Clinton Engineer Works derived its name, became a community in 1801. Shortly after the State's General Assembly approved an act taking enough territory from Knox County to form two new counties, Anderson, named for Joseph Anderson, then a U. S. Senator from Tennessee, and Roane, a commission composed of William Lea, Kinza Johnson, William Standifer, William Robertson, Joseph Greyson, Solomon Massengale and Hugh Montgomery was appointed to select a county seat for Anderson. The site chosen, it was named Burrville, in honor of Aaron Burr. When Burr fell into disrepute, the General Assembly of 1809 changed the name to Clinton in honor of DeWitt Clinton of New York.

In 1942, Anderson County had a population of 27,000; today it is 60,000. Clinton, with a population of 2765 in 1942, rose to 7,000 during the war years and now has a population of 3800. Census Bureau figures reveal that the dollar volume of retail sales in Anderson County increased 851 per cent from 1939 to 1948, a total of $29,700,000 in 1948 compared to $3,100,000 in 1939—a direct result of the building of Oak Ridge.

Scarboro, one of the three small communities in the Oak Ridge area in 1942, was founded in the late 1700's by three brothers from Virginia—Jonathan, David and James Scarboro. They were joined later by the Peters, Keith and England families and Scarboro flourished as a farming community until the fateful Fall of 1942.

Even the legend of gold pervades the Oak Ridge area. The story persists that $20,000 in gold coin remains hidden to this day in the vicinity of what is now Grove Center, one of the principal business sections of the new Oak Ridge.

Shortly before the Civil War, Collins Roberts drifted down from Connecticut, received a land grant of between 3,000 and 4,000 acres from the Government and began the development of a community which came to be known as Robertsville. There he built his home with the help of slaves. Just before the War Between the States broke out, Roberts decided to sell his slaves. In the transaction, he demanded gold, which he received in pieces of ten and twenty dollars.

He had hoped to keep the transaction quiet but the news spread that he had the money. He became so disturbed over the possibility of theft that he hid it. He disclosed the hiding place to no one. A short time later he died, carrying the secret of the gold to his grave.

The gold has never been found, although many years ago people came from all over the South, some with strange divining rods, camping out on a wide expanse of land, to search for the treasure. Where modern apartments now stand in Oak Ridge, mercenary persons once burned barns and other buildings searching for the treasure.

The treasure found at Oak Ridge was of another type.

U. S. Boom Town of the 1890's

Harriman, largest town in Roane County with a population of 7,000, once was America's Number One "boom town."

In the fabulous 1890's, a group of prohibition advocates led by General Clinton B. Fisk, who received half a million votes for President in 1888 on a Prohibition Ticket, determined to create an industrial city free of liquor traffic, a city in which a moral principle was to be successfully combined with a profitable commercial venture. Between 1890 and 1893, great hordes of people from the United States were attracted to the new city.

General Fisk and other Northern capitalists who formed the East Tennessee Land Company in 1889 with a capital of $3,000,000 envisioned an industrial city of half a million.

The group bought 275,000 acres of coal lands, 250,000 acres of farm and timber lands and leased or purchased all the known iron ore beds in East Tennessee; they foresaw growth of industry because of "rich iron beds" around the site, contending that enough coal to supply what they thought would be the largest steel center in the south was available in the nearby mountains.

Harriman also was to be a textile center, they claimed, chemical tests having proven that water from Emory River was the softest and purest in the United States. And virgin forests sufficient for operation of giant woodworking plants were nearby.

The Land Company, in addition to General Fisk, had among other officials W. C. Harriman, of Warner, N. H., son of Gen. W. C. Harriman, former governor of New Hampshire, for whom the town was named; I. K. Funk and A. W. Wagnalls, of the publishing firm of Funk and Wagnalls; Ferdinand Schumacher of Akron, Ohio; W. H. Russell, one-time owner of the Boston Braves baseball team; John Hopewell, Jr., and J. R. Leeson of Boston; A. A. Hopkins of Rochester, N. Y.; James B. Hobbs, Chicago; Francis W. Breed, Lynn, Mass.; William Silverwood, Baltimore; E. M. Goodall, Sanford, Maine, and Frederick Gates, New York.

Harriman was chartered in February 1890. First land sales began February 26, 1890, more than 5,000 men from 15 states attending. Before turning the city over to corporate authorities—the first city government was elected in June 1891, on a prohibition platform—the East Tennessee Land Company spent $50,000 building streets, an electric plant and waterworks, a school building and a meeting hall. By the end of 1891, there were 15 industries in Harriman. On January 1, 1892, the population was 3,672 and there were 27 corporations with a total capitalization of $7,335,000.

[40]

In 1893, the American University of Harriman was founded, chartered on the principles of temperance and prohibition. The co-educational college attracted students from 25 states. It soon passed into oblivion.

The Panic of 1893 ended the grandiose schemes for Harriman, envisioned as a Pittsburgh-on-the-Emory. The promoters did not return. The large coal deposits have never been worked out; the iron contains too much phosphorus for extensive use. One blast furnace still operates, however, in Rockwood, in Roane County. And a large hosiery mill, using the pure water of Emory River, is still operating in Harriman after 37 years.

CHAPTER IV

Of Those Who Came

WITH the establishment of the Manhattan District August 13, 1942, a time limit of three years was set for the development and use of atomic energy as a weapon of war. This objective was off schedule by only two weeks, an incredible achievement. The accomplishment in construction has been described as the equivalent of building a Panama Canal each year for three consecutive years.

Simultaneously with the selection of the Tennessee site, determination had been made to obtain uranium-235 for use in an atomic weapon through two methods, the gaseous diffusion and the electromagnetic processes. These plants were to be at Oak Ridge. A third plant for obtaining U-235, the thermal diffusion method, was authorized for Oak Ridge in 1944.

On December 2, 1942, in an experiment beneath the West Stands of Staff Field at the University of Chicago, a small group of scientists under the leadership of Dr. Enrico Fermi and Dr. Arthur H. Compton witnessed the advent of a new era as man first initiated a self-sustaining nuclear chain reaction and controlled it. Through the medium of such a chain reaction, production of plutonium, another element for use in an atomic weapon, was possible on a large scale.

[42]

Plutonium is a man-made element not occurring naturally on earth.

The successful step resulted in a decision of the Manhattan District to establish a plutonium pilot plant at Clinton Engineer Works. To carry out plutonium production on a massive scale, 400,000 acres near Pasco in the State of Washington were selected for construction of plants. This became known as the Hanford Engineer Works, now called the Hanford Works.

This installation cost $382,000,000 and was built and managed during wartime by E. I. du Pont de Nemours and Company. It is now operated by General Electric Company.

Early in 1943, the Manhattan Project established in New Mexico its most secret installation—Los Alamos. The site of 45,000 acres is a pine-dotted mesa towering nearly 7500 feet above sea level where young men once attended the exclusive Los Alamos Ranch School 35 miles northwest of Santa Fe, the state capital. This became headquarters for developing the techniques and mechanisms of atomic bombs using uranium-235 from Oak Ridge and plutonium from Hanford.[1]

---

[1] The atomic energy undertaking begins with uranium ores and ends with uranium-235 and plutonium, the fuels of atomic energy Uranium first acquired a commercial interest around the beginning of the Twentieth Century when radium was discovered and put to use. Radium always is associated with uranium, both coming from pitchblende, which is heavier than iron, about as hard as steel and is grayish black, sometimes with a greenish cast.

Major fields of pitchblende are in Canada and the Belgian Congo. They are the source of most high-grade uranium ores. Our own country has so far produced little high-grade uranium ores, although progress is being made in processing low-grade uranium-bearing metals, of which carnotite ore found in the Western United States is one.

Uranium ore is mined like any other element, gold, silver, copper, iron. The ore is then processed through large rolling mills, where it is crushed and refined. Reasonably pure uranium is then delivered to other plants within the United States for further chemical processing as feed material for the units

The course set, 1943 dawned over Tennessee amid a ca-
cophony of bulldozers, caterpillars and other earth-moving
equipment; the whirr and the whine of machines and saws
clearing the valleys and the ridges, the sounds of blasting and
the clicks of the time-pieces in clock-in alleys with the mount-

---

at Oak Ridge which produce uranium-235, and for those at Hanford, which
produce plutonium.

This feed material takes the form of a gaseous uranium compound (for the
Oak Ridge U-235 production plants) and solid pieces of pure uranium of certain size (for the Hanford plutonium production plants and for other uranium chain-reactors).

The amount of uranium-235 in natural uranium is very small—about one
part in 140. Even more serious than its scarcity is the difficulty of separating
it from the more abundant uranium-238, which makes up about 99 3 per cent
of the pure uranium Both U-235 and U-238 are atoms, or isotopes (from the
Greek "iso," same, and "tope," place) which go to make up the pure uranium;
that is, they do not differ chemically one from the other, but they do differ
very slightly in mass or weight Of the two atoms, U-235 and U-238, only
U-235 is suitable, because of the rules of nature, as atomic energy fuel. U-238
will not fission in the manner desired for a chain-reaction blast.

Three ways were developed at Oak Ridge in World War II to obtain the
rare U-235 from the more abundant U-238 These are physical and mechanical separation methods—like separating cream from milk. The thermal diffusion process used tremendous quantities of heat to bring about a separation
of U-235 from U-238. The electromagnetic separation process whirled uranium
atoms in large semi-circular arcs in a magnetic field. U-235 atoms followed a
slightly different path from that followed by the heavier U-238 atoms. U-235
and U-238 then were collected at different points at the end of the arc.

The third process, gaseous diffusion, has proven the most efficient and eco-
nomical and is now the major source of U-235 In this process, a gas, uranium
hexafluoride, is circulated and re-circulated under great pressure through barriers containing billions of holes much smaller than the point of a pin—each
one actually being only around two-millionths of an inch in diameter As the
gas is pressured through thousands of successive stages into these invisible
holes, U-235 goes one way, the heavier U-238 another At the end of the
"line," U-235 is collected.

An entirely different method, a combination transmutation-chemical separation process, was developed to obtain plutonium, the fissionable material
produced at Hanford Plutonium is not simply extracted or separated from
the feed material, which are slugs of pure uranium metal. It is a new element
created by nuclear fission—a transmutation of one basic element (uranium)

ing flow of construction workers who were to reach a peak of 47,000 in Oak Ridge in the Spring of 1944 shortly before D-Day on the beaches of France. The peak operating force was 40,000 in May 1945. The overall peak employment on the project, construction, operating and other, was 82,000 in May

---

into another (plutonium), which is then chemically separated from the parent uranium, since it is a different chemical element. The transmutation occurs in huge reactors, or so-called atomic "furnaces" A substance far more precious and useful than gold is produced by this modern alchemy. Once produced, plutonium also has properties of fission, or breaking up, under certain conditions, with release of great amounts of energy.

One type of a plutonium-producing reactor is a huge "pile," or solid mass, of graphite pierced at intervals by tubes that run from one side of the pile to the other. Uranium, in the form of slugs, is placed in these tubes in certain geometric designs. Nuclear fission then transmutes a small portion of the uranium into plutonium

Chain reactions occur in both chain reactors, or "piles," and in atomic explosions In a "pile" operation, the reaction is controlled by skilled operators; in an atomic explosion, the reaction is uncontrolled, and runs its course.

Both U-235 and plutonium, under certain man-made conditions, will fission —that is, break up and actually lose part of their mass with release of great amounts of energy in a chain reaction, such as in an atomic bomb. Fission is a particular kind of disintegration of an atomic nucleus and an explosion can be produced by bringing sufficiently large masses of fissionable material together rapidly. An atomic weapon is a device for doing this.

A chain reaction is any chemical or nuclear transmutation in which some of the products of a particular change assist the continuation of that change. In the atomic bomb, or the power-producing uranium "pile," fission is caused by the capture of a neutron by a uranium atom. Then, when fission occurs, more neutrons are released, which in turn produce fission in additional uranium atoms, and so on. Thus a chain reaction. A neutron is a particle with no electric charge. Because of that, neutrons can move rather freely through solid matter.

Atomic energy itself differs radically from ordinary types of energy since it involves a fundamental change in the atom's nature. In this change, some matter is converted into energy. In burning coal, for example, carbon, hydrogen and oxygen atoms are regrouped into new molecules forming new substances. The atoms remain unchanged—they are still carbon, hydrogen and oxygen They do not lose mass of any consequence.

In releasing atomic energy, however, through splitting or breaking up of the atom, the atom changes identity completely. It loses part of its mass, which

[45]

1945. A total of 110,000 construction workers were hired from 1942 through 1945 to build the town and plants; around 400,000 construction workers were interviewed with only one out of three being employed.

Those who came in a steady stream in 1943 as the pace of building mounted worked amid scenes having all the frontier trappings of a Western movie. The slow-paced voices of Southerners mingled with the twang of the Midwesterner, the sharp, direct speech of the Easterner and the deliberate, color-

---

is converted into energy. The energy liberated is in proportion to the amount of atomic mass destroyed.

Elements, the basic substance of the universe, make up all matter. Each element is made up of atoms which are alike in chemical behavior but may have different weights. Furthermore, the atoms of each element are structurally different from atoms of other elements. Virtually all of the 98 elements discovered by mankind to date contain some types of atoms which undergo spontaneous "atomic explosions," but all such "explosions" in atoms of elements below uranium, element 92, are very minute when compared to uranium-235 or plutonium as they undergo fission, or break up.

The difference is one of degree, in that some of the atoms of the common elements break up so slowly over so many thousands and millions of years that the energy released is practically negligible. It is this slow, almost constant, rate of breaking up of all the elements in nature, plus the cosmic radiation from the sun, which is commonly referred to as background radiation (that radiation found to be normally present in any given geographical location). Background radiation in any location is dependent on the kinds and abundance of the elements present in that location. Mankind is constantly being bathed in background radiation, it is part of our daily lives.

Why, one may ask, cannot fission of major proportions as that occurring in uranium-235 or plutonium be accomplished with atoms of ordinary stable elements—tin, zinc, aluminum, sodium, copper, iron?

The answer, simply stated, is that the "excitement" or "energy level" within the atoms of common elements is too low. When uranium-235 or plutonium soaks up a neutron, a very high degree of "excitement," or energy level, is reached. To relieve this excitement, a major change in the atom occurs, such as the splitting into two more or less equal parts and the release of a great deal of energy. When atoms of the more common elements take in a neutron, the same high degree of excitement is *not* reached, and the energy is released in a less spectacular and much more deliberate manner. In other words, the nuclei of fissionable materials must have a great amount of excess energy.

[46]

ful talk of the Westerner. License plates of the thousands of automobiles (25,000 daily at one time) which poured into the project and ofttimes ran virtually bumper to bumper during the mornings and the afternoons between the project and Knoxville, Clinton, Lenoir City, Maryville, Harriman, Oliver Springs, Rockwood, Lake City, Kingston and other nearby communities, were from every state in the Union.

A cosmopolitan city which was to be 7 miles long and from 1½ to 2 miles wide, approximately 9,000 acres or 14 square miles, began to take roots in the northeast corner of the area; in once isolated valleys, foundations for great new structures were beginning to take form.

The original builders and workers were drawn from every stratum of American life. From Pittsburgh came the iron and steel workers; from Grand Rapids the woodworkers; from Detroit the machinists; from the TVA the electrical experts. Truckers, riveters, crane operators, carpenters and other craftsmen, and clerks, stenographers, auditors, accountants and general help came from everywhere. The little and the big—the laborer to the scientist—worked in a common cause.

Their sacrifices, their perseverance, their will to get a job done rivalled the resoluteness and strength of the pioneers who had come into the same valleys 150 years before. The newcomers, 65 per cent of them from the states comprising the great Tennessee Valley and adjoining states in the South, came for the purpose of building a city with a purpose, although at the time they knew not why.

Simultaneously with the surge of people into the new area, stores, shops and various other commercial enterprises moved into hastily-assembled structures—awaiting the building of more permanent facilities—to serve the needs of the newcomers; prominent in the vanguard was the famous Fuller

Brush "man," his wares prominently displayed, with an appropriate sign, in front of a "hut" in the midst of the thriving new business section.

Those who came learned to live in deep mystery, isolating themselves along with the project, all the while hoping they were making some tangible contribution to the war effort. They, too, were following an uncharted path. Many felt as the old Negro, who, showing his identification badge as a laborer, requested a bus driver to "please, suh, look at my badge and tell me where I want to go."

All in all, they came for patriotism. Between January 1943 and June 1945, 5500 persons in various occupations were deferred from the Armed Forces for work with the Manhattan District, at Oak Ridge and elsewhere. And the tough Americans, those who build the skylines of America, came because they like to beat tough assignments. Oak Ridge enticed them all.

Site preparation for the community began in October 1942 immediately after Stone & Webster construction personnel reached the site. The first structure begun was the main administration building of the Manhattan District which was started November 22, 1942, and completed March 15, 1943, enabling key persons of the District to move from New York. Meanwhile, in Washington, War Production Board officials became familiar with the strange words "Manhattan District." Throughout the building of Oak Ridge and other atomic facilities, the WPB gave valuable assistance in obtaining vital supplies and equipment for a project about which they knew little.

The first phase of Oak Ridge community planning evolved early in 1943 for a town of around 13,000 residents. The second phase, initiated in the Fall of 1943, provided for a population of around 42,000 and the third phase, initiated in the Spring of 1945, provided for a population of 66,000. Even this latter estimate proved too low; the peak population of the Oak Ridge Area in September 1945 was 75,000 (the fifth largest city in Tennessee, exceeded only by Memphis, Nashville, Knoxville and Chattanooga), occupying 10,-000 family dwelling units, 13,000 dormitory spaces, 5,000 trailers and more than 16,000 hutment and barracks accommodations. In the 1950 Census, Oak Ridge had a population of 30,205, maintaining its place as Tennessee's fifth largest city.

During one period of the construction of the family dwellings known as Cemestos, units were turned over to the Government by the contractors at the average rate of one every thirty minutes—fully equipped with refrigerators, stoves, heating equipment and all the facilities necessary for immediate use.

The first of the Cemesto, or semi-permanent-type houses, built under supervision of Stone & Webster, were constructed south of Tennessee Avenue—six months before only a field dotted with pines—in what was designated the Elm Grove Community. All the foundations in a given area were laid at once, the chimneys coming next—the red brick looking like strange trees—and then the Cemesto siding was slipped into place.

In July, prefabricated wooden huts were ready for occupancy by construction workers and trailers were obtained from the Federal Public Housing Authority and wheeled into place at Middletown. Looking across from the Manhattan District's main administration building at night in July, few patches of light designating houses could be seen. But week by week, they spread on and on and Black Oak Ridge sparkled like so many stars.

By August, plans for the project were revised upward.

[49]

More housing units were needed for the construction forces which were simultaneously working on the huge plants four, 10 and 12 miles away and for the increasing number of operating and technical personnel who were beginning to move in. To save time and labor, prefabricated houses were brought in by trucks, clogging the highways for months. Panélized duplex houses constructed at other war plants were knocked down, shipped to the area and reassembled. Two and four-family units and 12-family apartments were built. More Government trailers were obtained and accommodations provided for privately-owned trailers. Along with the housing, schools, stores, recreation and general facilities were enlarged accordingly.

The first family moved into its trailer home in Middletown on July 13, and on July 27, 1943, the first Cemesto house was occupied on Thornton Road, off Tennessee Avenue. In the same month, Col. Robert C. Blair, then in charge of Clinton Engineer Works, asked for suggestions from employees for a name for the new community. Oak Ridge was chosen because the site on which the town was being built was known as Black Oak Ridge and its rural connotation held outside curiosity to a minimum.

Earlier, in January of 1943, ground had been broken for the first fire headquarters, for the first dormitory, and for recreation buildings. On February 1, construction of the great electromagnetic plant was begun. On the same day, grading was begun at the site for Clinton Laboratories, site of the first uranium chain-reactor which was to supply the initial significant amounts of plutonium for research leading to the building of the great Hanford plants. Rigid control over project entry was established April 1, 1943, when armed guards were placed at "gates" at Elza, Edgemoor, Blair, Solway and other road points. Wire fencing began to appear at strategic places and mounted patrolmen began to watch the stretches of Clinch River which meanders 35 miles around the project site.

The family of Capt. Philip Anderson of Little Rock, Arkansas, an officer assigned to the Manhattan District, was the first to move into a semi-permanent-type home in Oak Ridge. Like thousands of other families who came later, the Andersons underwent experiences seldom seen since the Gold Rush.

"It was raining the day we moved in," Mrs. Anderson recalls, "and the truck with the furniture got stuck in the mud. When we finally struggled into the house, the painters were still in the kitchen. There were no phones in the houses at that time and when they had trouble with the electricity they had no way of letting you know. I'd be in the middle of getting dinner ready for company, when the lights and the stove would go off, twice when I was washing my hair the water went off. . . . We weren't born soon enough for the Gold Rush, but we made up for it at Oak Ridge."

The new city's first drug store opened August 1, the first grocery store made its first sale August 4 and the Guest House, the local hotel, welcomed its first guest August 5. This wooden structure was to serve as a stop-over for many of the world's leading scientists and executives—among them Dr. J. Robert Oppenheimer, Dr. Ernest O. Lawrence, Dr. Harold Urey, Dr. G. T. Seaborg, Dr. Enrico Fermi, Dr. J. B. Conant, Dr. George T. Felbeck, H. D. Kinsey, Crawford H. Greenewalt, James A. Rafferty, T. J. Hargrave, James C. White, P. S. Wilcox, P. C. Keith, Jr., Dr. Eugene Wigner, Dr. Vannevar Bush, Dr. Charles A. Thomas, Dr. W. H. Zinn, Dr. Arthur H. Compton and the Englishman, Dr. M. L. Oliphant—all moving like shadows through the mysterious recesses of the plants, unrecognized by the thousands who were building what they, the scientists, had planned. For

security reasons the scientists went under assumed names—
Dr. Fermi, for example, was "Farmer," Dr. Compton "Co-
mus" and Lawrence "Lawson."

The first movie "In Which We Serve" was shown in the
Center Theatre August 31, fifteen days after the U. S. Post
Office was opened by the first postmaster, George E. Bowling,
and the first letter was dispatched bearing the Oak Ridge
cancellation stamp.

"We opened the first grocery and made our first sale even
before the construction crew got the doors up," Horace
Sherrod, the manager, states. "I started filling my shelves in
1943 and didn't manage to get them completely filled until
1946. When I asked the big shippers to send me food, they'd
say 'We never heard of Oak Ridge, it can't be a priority city.'
A carload of merchandise I ordered hadn't arrived in four
weeks and I inquired about it. The shipper wrote back and
said nobody could tell him where Oak Ridge was; he said he
didn't believe there was such a place. A flour salesman lost
hundreds of dollars in commissions because his company
couldn't locate Oak Ridge after the orders arrived, and they
refused to ship."

In the Summer of 1943, Oak Ridgers began looking for
recreation. The residents needed some place to shake the mud
off their feet and the grime off their hands. With the completion of the first cafeteria, this was utilized for dances under
the direction of the Oak Ridge Recreation and Welfare Association, which was activated July 21 and which eventually had
230 full-time employees supervising social, welfare and recrea-
tional activities encompassing a "teen-age" center, athletics,
library services, folk dancing, music, art, drama, handicraft,
outdoor dancing and other diversified undertakings. The
association was dissolved in 1947.

The first issue of a small four-page mimeographed publication known as the *Oak Ridge Journal* (it eventually grew to a regular size newspaper of 12 pages under the editorship of Mrs. Frances Smith Gates, a graduate of Mt. Holyoke College, assisted by Richard B. Gehman, Dixon Johnson and Thomas F. X. McCarthy), came to the attention of residents September 4 with Army Warrant Officer Murray Levine of Brooklyn as editor. He was succeeded a short time later by Carl Jealous. The *Journal's* second issue called for the organization of clubs and associations. The reception was enthusiastic. By the Summer of 1945, Oak Ridge had more than 100 organized groups ranging from a Rabbit Breeders Association to a Folk Dancing Group to a Civic Music Association. The first civic organization in Oak Ridge, the Junior Chamber of Commerce, was formed in May, 1945; the Rotary Club was second, being organized in December, 1945. The Lions, Kiwanis and Exchange Clubs followed. The Masons, Elks, Eagles and Moose and other groups developed Oak Ridge organizations at intervals.

On September 12, the Interdenominational Young People's Fellowship Union staged a campfire sing at a pond close by the new community. A Civil Air Patrol was organized September 14 and plans were well underway for a full-fledged Red Cross chapter. Army enlisted men and members of the Woman's Army Corps began arriving. On September 18, Boy Scouts were signing up for organizations of Troops and a Little Theatre was formed. Bowlers organized an association on September 30 with 86 men's and 27 women's teams.

On September 23, the Army announced a contract with an organization which was to become a close part of the daily life of every Oak Ridge resident—the Roane-Anderson Company, the housekeeper for the Atomic City (Chapter VII).

The frontier atmosphere of Oak Ridge's early days is reflected in a letter to the residents from the first Town Manager, Capt. P. E. O'Meara of the Manhattan District, appearing in the mimeographed *Oak Ridge Journal* September 25, 1943. The letter said:

"Yes, we know it's muddy. . . . You think prices are too high in the grocery store. . . . Coal has not been delivered. . . . It takes six days to get your laundry. . . . The grocer runs out of butter and milk. . . . Your laundry gets lost. . . . The post office is too small. . . . There are not enough bowling alleys. . . . Your house leaks. . . . Everyone is not courteous. . . . It takes too long to get your passes. . . . The water was cold. . . . The beer ran out. . . .

"The telephones are always busy. . . . You can't get all the meat you want. . . . Your house isn't ready. . . . There's confusion in the cafeteria. . . . The dance hall is crowded. . . . There's no soda fountain. . . . The guest house is full. . . . Employees are inexperienced. . . . You don't like the way things are run. . . . You could do better. . . . Someone said someone asked someone who told them someone said they knew something, and you don't like it. . . . Your windows aren't clean in your house. . . . You have seen the movie. . . . Your floors aren't waxed. . . . The butcher didn't wait on you in turn. . . . You want more sugar. . . . The roads are dusty. . . . Your shirts come back without buttons. . . . Things were different 'back home'. . . . You would have planned it differently. . . .

"What you want to know is . . . WHAT'S BEING DONE ABOUT IT?

"Well . . . Roads WILL be paved. . . . The grocer is obligated to not charge prices in excess of those in Knoxville. . . . Coal WILL be delivered. . . . Sidewalks WILL be laid. . . . A third shift will be started in the laundry as soon as we can get help. . . . Milk WILL be imported, maybe butter. . . . The townsite WILL be restricted. . . . An officer is in Washington now arranging for the change from a fourth to second rate post office. . . . More bowling alleys WILL be built. . . . Workmen WILL come by and ask where your house leaks.

"Town Management personnel has been instructed that YOU are always right. . . . Personnel estimates increased faster than dorms could be built, more WILL be built. . . . They ran out of beer in Knoxville and 'back home' the same night it ran out here. . . . More telephones are coming. . . . Meat is rationed. . . . 3,000 people cannot be fed in two hours and not have confusion.

"Every effort is being made to get your houses ready. . . . Construction must go on, even when you are asleep. . . . More dance space WILL be made available. . . . The guest house will be full for months. . . . Soda fountain equipment is almost unobtainable, but we WILL have one. . . .

"Some employees will ALWAYS be inefficient. . . . Someone will always be saying someone asked someone who told them someone said they knew something, that DOESN'T make it a fact. . . .

"We WILL get more first run movies. . . . A plan is ready to furnish clothes line posts and tools and you put them up. . . . The butcher didn't mean to pass you up. . . . The barbers are busy back home during rush hours. . . . Sugar is rationed. . . . Your shirts will continue to come back without buttons; we would put them on if we had help enough. . . . A shoe repair shop WILL be opened soon. . . .

"Your dormitory WOULDN'T be noisy if everyone were as considerate as he would like his neighbor to be. . . . Were you ever ANYWHERE that you liked everyone. . . . Things WEREN'T different back home. . . . Everything can't be done at once, because we need more help. We would have planned it differently too if we had thought of it in '33."

Early in October, 1943, Oak Ridge's first school sessions,

where the engineer's boy from Nebraska sat next to the accountant's son from Brooklyn, got underway.

The Oak Ridge Hospital was completed later in the month, with formal opening November 17, 1943. The first staff members under the supervision of Dr. Charles Rea, a Minnesotan and a Colonel in the Army Medical Corps, had arrived in July. The hospital was under the direction of Col. Stafford L. Warren, chief of the Medical Division of the Manhattan District, with Lt. Col. H. L. Friedell as Executive Officer. Colonel Rea was director of clinical services.

With construction on all sides, the dust created a major problem in hospital operations and the job of keeping operating room instruments sterile was a major difficulty. Between July and October, while the hospital was nearing completion, the hospital staff conducted what was almost a mobile hospital unit, setting up clinics and giving typhoid and smallpox inoculations in the cafeterias. A key figure at the hospital was a psychiatrist, for the mental health of the worker was of prime concern. The long, critical hours, the elaborate security and safety rules, the unreality and the need to produce at top efficiency and speed brought about pressure of great magnitude.

In November 1943, the hospital's bed capacity was 50; by the summer of 1945, it had grown to 337; today 260 beds are utilized. The hospital, now operated by a nonprofit corporation composed of seven representative Oak Ridge citizens, is a member of the American Hospital Association and is approved by the American College of Surgeons and the American Medical Association for intern training. A completely equipped dental clinic is nearby.

House-to-house milk delivery also began in October, 1943. Housewives with children were the first to get delivery. Condensed milk, recombined with water, augmented the local

From these early scenes (the site in the bottom photo is now Jackson Square, business section) . . .

Manhattan District

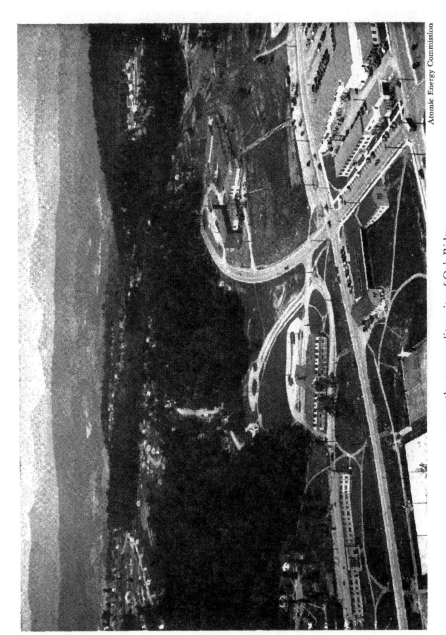

. . . grew the cosmopolitan city of Oak Ridge . . . .

. . . with a peak population in 1945 of 75,000 (hospital in foreground) . . .

together with schools

. . . and churches (Chapel-on-the-Hill) to serve its residents . . .

. . . and neat, attractive homes

supply and milk sheds as far away as Wisconsin were contacted for supplies. Even then, there was no laboratory examination of milk sold on the area—most of it came in bulk in large cans—until January 1944, when the Department of Public Health was set up. During the building of Oak Ridge and in peacetime, the community has established an enviable health record. There has never been a serious outbreak of any disease or an epidemic in Oak Ridge.

October also saw the establishment of Oak Ridge's first banking facility, an extension of the Hamilton National Bank of Knoxville. On November 18, 1942, the first telephone was installed on the project by Southern Bell; in 1945, the total was almost 10,000. Today, approximately 14,000 telephones serve Oak Ridge.

The women who came to Oak Ridge with their husbands to help share the pioneering in a new land were distressed over what they found. Security demanded silence and silence was golden in Oak Ridge. The wives were upset when they first saw the drab countryside of Oak Ridge. But they learned to take it in stride. This transition was aptly described by a young stenographer, a dormitory resident: "You either catch on here fast, or you don't catch on at all."

Even fun had an awkward time. Young girls, immaculately dressed for dances, often removed a mud-caked pair of hip boots at the door, stacked them in racks in the hall with others and then slipped on their dancing shoes. Furthermore, it was considered proper to remove one's shoes before entering a house.

As related by Mrs. Stafford Warren, a columnist for the *Oak Ridge Journal*, housewives confined their conversations to lighter topics and were discreet in inquiring as to their husbands' destinations on trips. They got used to a "left-out" feeling. For the wives, 1943 through a part of 1945 was a two-

and-a-half-year period of a concentration of curiosity. In the "outside" world, relatives and friends also had their doubts about Oak Ridge.

"It's all right, son," one mother wrote, "if you can't tell me what you're doing in Oak Ridge, but I do hope it's honorable." A friend in another city wrote to a housewife: "For a bright young engineer, your husband certainly knows less about what he is doing than I would have expected" and a relative in Chicago put it this way: "If you and John don't care to correspond with your old aunt any longer, just let me know."

On the other hand, those on the outside couldn't properly sympathize with Oak Ridgers on the agony of shopping for food in the early days because the new resident couldn't explain that one grocery store was attempting to serve 10,000 persons, nor could they appreciate why church members were not chagrined at the sign over an Oak Ridge movie theatre entrance: "NOW SHOWING: METHODIST CHURCH." The church was happy to have the accommodations.

One scientist worked out his own technique for avoiding discussions with his wife on the nature of his work. He said to her: "I'll give you a choice. I'll tell you exactly what I'm doing, but then you can't breathe a word of it to a soul. Or I'll continue to keep you in the dark and you can make up all the stories you want to tell your friends." She thought it over, and being a woman, took the choice of not knowing and being able to speculate.

During the hectic period when Oak Ridge was born, tempers flared and tensions mounted in the wake of the thousands upon thousands of persons who swelled the populations of Knoxville and adjoining communities during rationing and tightening of the belts. Quarrels developed between the natives and the newcomers. The Knox County Court—made up of "squires"—adopted, 22 to 3, a resolution seeking a Congressional investigation of reports that some Clinton Engineer Works' employees were paid for work they did not perform. The action was dropped but it was a reflection of the times.

The confusion was contagious. One workman, on the job only five days, insisted to his superior that he wanted to "terminate." The superior asked why he was dissatisfied.

"Well," the worker said, "it's that paper they wanted us to sign, sign up for the 'fire squad.' I'm willing to do all I can here but I don't like killing Americans. I don't want to have to fire on anybody in Oak Ridge." He was in the same mental state as the Negro who called the Oak Ridge Hospital: "Please send an ambulance to the colored huts; this man's havin' a confusion" and the Mounted Patrolman who aimed at a skunk and shot a perfectly good Government-owned horse through the head.

At 5 A.M. on the morning of November 4, 1943, a historic first step in Oak Ridge operations was taken. At that time, the first uranium chain-reactor in the world with a production potential was placed in operation, the famous atomic "furnace" at Clinton Laboratories, to be known later in peacetime as the Oak Ridge National Laboratory. The "pile" had been completed in eight and one-half months by E. I. du Pont de Nemours and Co. and turned over on October 16 to the Metallurgical Laboratory of the University of Chicago, which originally operated the completed plant under the leadership of Dr. A. H. Compton.

Meanwhile, construction of the main process building for the gaseous diffusion U-235 plant was started September 10 by the J. A. Jones Construction Company of Charlotte, N. C.

Oak Ridge's first Christmas in 1943 was a painful one for most families. They were away from relatives and old ties and Security did not encourage visitors. Work went on apace, for the loss of only one day in construction meant a one-day delay sometime in the future in the use of mankind's newest weapon. And the children did not ask Santa Claus to bring the usual toys. They were interested in what they lived around—bulldozers, dump trucks, caterpillars, wheelbarrows and cranes.

Oak Ridge was a city of boardwalks, badges and some bootleggers but never boredom. There was a touch of the Klondike, the fervor of a boom. Friends one hadn't seen in years would suddenly appear and as suddenly disappear. Scientists and mountaineers worked side by side, but the only people really at home were engineers who had experienced construction before.

Early in 1944, Stone and Webster completed the first operations building in the production area known as Y-12 and on January 27 a select group of Manhattan District personnel and officials of Stone and Webster and the Tennessee Eastman Corporation, a subsidiary of Eastman Kodak and plant operator, witnessed the epochal first "run" of uranium-235 on a mass basis by the electromagnetic method. The run was marred by discovery of foreign matter in certain pipes which had to be immaculately clean for successful operations, but the ingenuity of engineers and technical personnel quickly solved the problem and in February, first full-time production was under way. The uncharted path was beginning to clear.

Few Oak Ridgers were aware of an event transpiring in 1943 which was to affect each and every resident.

Early that year, the Tennessee Legislature, alarmed by the growing number of Federal projects in the State and the subsequent loss of state lands and taxes from such lands, declined to cede sovereignty to the Federal government over the land taken over by the War Department for the Oak Ridge project.

Prentice Cooper of Shelbyville was Governor of Tennessee at the time.

Thus, Oak Ridge is not a Federal Reservation, merely a Federal area, and the people who lived there during the war years and who live there now were and are beholden to the Criminal and Civil laws of the counties of Roane and Anderson and the State of Tennessee. Until the community becomes an incorporated municipality of the state, laws will be enforced by Oak Ridge policemen deputized by the sheriff of Anderson County.

As residents, however, of an area in which state laws prevail, the citizens, after residence in Tennessee of a year, can vote in state and county elections. The decision of the Legislature in 1943 not to cede sovereignty eventually brought a healthier condition.

When peace came and Oak Ridge started on the road to normalcy, the transition in becoming a more integral part of Tennessee's civic, business, political and community life was facilitated.

Cost and Layout of Oak Ridge

The cost of building the town of Oak Ridge alone was $96,-000,000.

Its construction as a service to the atomic energy plants was accomplished by Stone and Webster Engineering Corporation to provide living quarters for personnel necessary for plant operations.

Also assigned the construction of the huge electromagnetic facility, Stone and Webster turned over the town planning and housing phase to the architect-engineer firm of Skidmore, Owings and Merrill of Chicago, which prepared building plans and town layout. The John B. Pierce Foundation of

Raritan, New Jersey, developed the designs for semi-permanent housing units.

Stone and Webster was responsible, however, for coordination of all work, procurement of materials, supervision of sub-contractors building the homes, and construction of roads and utilities. Principal sub-contractors in building the town were John A. Johnson & Sons, Inc., New York; Clinton Home Builders, Charlotte, N. C.; O'Driscoll and Grove, Inc., New York; Harrison Construction Co., Pittsburgh and Maryville, Tenn.; Foster and Creighton Co., Nashville; A. Farnell Blair, Decatur, Ga., and Rock City Construction Co., Nashville.

The layout of the nearly 200 miles of streets for the city was controlled mainly by the contours or grades of the area. Thus there are no "blocks" or "squares" as in other cities.

Several main roadways run east and west. Oak Ridge Turnpike was once the southern limits but new neighborhoods have now been developed south of the Turnpike. Tennessee Avenue runs through the main section of the town. Continuations to the west of Tennessee Avenue are Pennsylvania, Hillside Road and Robertsville Road. Outer Drive runs near the top of the ridge.

Main "Avenues," which generally run north and south or up and down hill, connect these three main east and west roadways.

Roads, Lanes and Places branch off the Avenues. These branches are called "Roads" when they form connections between other roadways; "Circles" when they form short loops which return to the same road from which they started, and "Lanes" or "Places" when they have dead ends.

Except for the main east and west connecting roads, all Avenues of the town as originally laid out have names with first letters progressing alphabetically from east to west. For example, "Alabama" and "Austin" on the east, "Georgia" to "Kentucky" near the center, and "Vermont" and "Victoria" toward the west. Later expansion of the town, west of Pennsylvania Avenue, started with the street names beginning with "H," "I" and "J," and continued alphabetically to the west with other letters not used in the original half of town.

Similarly, the names of "Roads," "Lanes" and "Circles" progress alphabetically from south to north or east to west, but with the same first letter as in the name of the "Avenue" from which they start. Thus all Roads and Lanes leading from Florida Avenue begin with "F," while all which lead from Outer Drive begin with "O."

Highest point above sea level in Oak Ridge is 1204 feet on Outer Drive; lowest is 775 feet along Clinch River, which serves as an area boundary.

The average mean temperature in the Oak Ridge Area over a period of the past 78 years has been approximately 60 degrees.

Schools, Churches and Babies

First school sessions in Oak Ridge began October 4, 1943, with an enrollment of 830 under the direction of Dr. A. H. Blankenship of Columbia University, superintendent, who arrived in July 1943 to build a school system which was to accommodate pupils from every state in the Union and from every conceivable background and which later was to be chosen one of the 40 most modern school systems in the nation. High point in Oak Ridge school enrollment was 8,223 on October 5, 1945; present day enrollment approximates 6,300. Dr. Blankenship left Oak Ridge in 1946 and was succeeded by Dr. R. H. Ostrander January 1, 1947.

First baccalaureate service in Oak Ridge school history

was held June 18, 1944, in the high school auditorium. Invocation and benediction were given by the Rev. W. A. Rule of the Baptist Church; the sermon was delivered by Father Joseph H. Seiner of the Catholic Church. On June 20, the first senior class was graduated from the high school. Dr. Blankenship presented the diplomas. The invocation was given by the Rev. B. M. Larson. Carolyn Lakin and Stafford Warren tied for first honors; second honors went to Barbara Lee Wensel.

There were 16 churches in the Oak Ridge Area when it was selected for building of atomic energy plants in 1942. Of these, five were in the section where the town was built and they were used by new congregations made up of the families of men who came to build and operate the town and plants. Church bells were removed from those remote from the residential section and installed in fire headquarters throughout the area.

Two new churches, Chapel-on-the-Hill in the central part of the city and another in East Village, East Chapel, were built by the Army. Two of the structures already in the Townsite area in 1942, West Chapel and Iroquois Chapel, were set aside for services but later it was necessary to utilize the high school auditorium, the grade schools, recreation halls and theatres for the 29 different church groups which eventually were established at Oak Ridge.

The first formal church services were conducted in Oak Ridge July 25, 1943, in the Central Cafeteria by the Rev. B. M. Larson, a Presbyterian, of Knoxville at the request of a United Church group totaling 154. Father Joseph H. Seiner said the first Catholic Mass in Central Recreation Hall August 22, 1943, with 85 attending. As the population of Oak Ridge increased, other denominations organized and by the time Chapel-on-the-Hill was dedicated October 19, 1943, the Baptist, Episcopal, Methodist, Jewish and other congregations were prepared to conduct services.

In the present-day Oak Ridge, church groups are building their own churches on land purchased from the Government.

The first child born in the Oak Ridge Hospital was Robert William O'Neal, son of Mr. and Mrs. James O'Neal, at 8:34 o'clock on the night of November 11, 1943, six days before the hospital was formally opened. Since that time, more than 6,700 children have been born in the hospital.

Birth rates per 1000 population in Oak Ridge generally exceed Tennessee and U. S. rates. In 1948, for example, Oak Ridge rates were 27.4 per 1000 population; Tennessee was 26.4 and the U. S. 24.2.

Death rates per 1000 population in Oak Ridge for 1948 were 2.7 as compared to 9.4 for Tennessee and a slightly higher figure for the U. S.

CHAPTER V

They Couldn't Say a Word

THE late John Sharp Williams, United States Senator from Mississippi, once appeared on the same platform with a tiresome opponent, who, when he made his speech, filled it with such expressions as "Dame Rumor says that Senator Williams, etc.," and "Dame Rumor hath it that Senator Williams, etc." On completion of his opponent's address, Senator Williams would arise and say "Dame Rumor lies" and then sit down.

Those in charge of building Oak Ridge found Dame Rumor operating at peak efficiency in 1943, 1944 and part of 1945. So irritating and at times threatening to the completion of certain operations did the rumors and gossip become that in 1944 the Manhattan District ordered a full-dress publicity campaign to arouse the South to the danger of loose references to operations at Clinton Engineer Works.

The campaign was directed at such remarks as:

"Thousands of kegs of nails are stacked up, run over by caterpillar tractors and covered up with bulldozers";

"Buildings erected according to blueprints, complete with expensive plumbing fixtures have been faced the wrong way, then saturated with gasoline and burned to the ground so as to hide the mistake from the general public";

[66]

"Wire sufficient to put a 15-foot fence around the City of Knoxville, but never unrolled, has been dumped into ditches and covered over";

"Hinges costing $70 per door are put in many buildings; similar practices in construction are followed";

"Millions of feet of first-class lumber have been saturated with kerosene and gasoline and burned up"; and

"A Social Security project of unparalleled waste not necessary to the war effort is being constructed."

The last charge was particularly obnoxious to the few who knew the project's objective since President Roosevelt had approved in the Summer of 1942 a $2,000,000,000 gamble on the word of several eminent scientists that they believed atomic energy could be developed on a large scale for use as a weapon of war. But among the rumors which continued to envelop the project were such as these·

"It's a Roosevelt boondoggle merely to have men at work," and

"It will be a home and a place for returning service men to live and work."

One observer has remarked that the only mystery to native Tennesseans was not the project itself but the mystery of why it was a mystery. He concluded that because they could see nothing tangible coming out of the smokeless, mysterious plants, even after operations had begun, they formed their judgments on a few fragmentary details. "They couldn't understand," he said, "why their Government just didn't announce in the beginning what it was trying to do and then proceed to do it, without all the fuss of secrecy."

But in a special memorandum from the White House in the Fall of 1942, President Roosevelt told General Leslie R. Groves, the Commanding General of the Manhattan Project, that "secrecy and security" were to be paramount. To carry out that mandate, there was set up in the Manhattan District an Intelligence and Security Division which was later described as "having successfully accomplished the outstanding security mission of the war."

This group which clothed and guarded with extraordinary care the objective of the project was comprised of chemical, mechanical, civil, electrical and fire protection engineers, physicists, lawyers, accountants, school teachers, public officers, newspapermen and linguists. The group operated independently of the Army Intelligence Organization but in close cooperation with it and the Federal Bureau of Investigation.

Oak Ridge was headquarters for the special detachment which carried out various assignments throughout the country under the direction from Washington of two men handpicked for the Project from the Army Military Intelligence Division—Col. John Lansdale, Jr., of Cleveland, Ohio and Col. W. A. Consodine of Newark, N. J. In Oak Ridge the detachment was first under the leadership of Major H. K. Calvert of Oklahoma City, Oklahoma, assigned to Col. K. D. Nichols, district engineer of the Manhattan District. Calvert later was sent to London to handle foreign aspects of the bomb development and was replaced by Lt. Col. W. B. Parsons of Seattle, Washington, with Capt. B. W. Menke as Executive Officer.

In maintaining secrecy, Manhattan Project agents traced rumors and speculations to their source. No serious sabotage occurred; what did happen could be attributed to personal spite. But espionage always was in the air. So well was the secret guarded, however, that the Germans believed, even to the war's end, that the United States had not progressed beyond the early research stages in development of the atomic bomb. With success of the project dependent upon the safety of key scientists, elaborate measures to protect their identity and whereabouts were carried out and agents who were safe automobile drivers and who would never lose a briefcase were assigned as constant bodyguards.

The agents went half way around the earth to check on and plug leaks. They went to the front lines in Belgium (an American officer wrote a friend in Oak Ridge asking if he were working on atomic energy), to France and the British Isles and to the uranium ore fields of the Belgian Congo and Canada They even chased a case of skin rash to South America; a workman formerly employed at Oak Ridge who may have been allergic to some common acid or to Tennessee poison ivy was rumored to believe his rash was from queer rays.

Their work carried them into many different fields. Stocks and bonds to be issued on the great exchanges by companies having contracts with the Manhattan District had to be worded so as not to reveal the atom secret. Education of thousands upon thousands of workers in how to talk and act with outsiders was necessary, as was the reading of hundreds of newspapers,.the scanning of science reports and attendance at scientific meetings to see there were no inadvertent slips of the tongue. They even examined books in libraries to determine whether sections on uranium and atomic energy had been thumbed too frequently. They fingerprinted over 300,-000 persons who came to work on the project.

They even denied a Bible salesman entrance to Oak Ridge in the Spring of 1944 because indiscriminate peddling was not allowed, and in Washington, Superman of the comic strips met his match when the Office of Censorship asked Superman's creator to delete mention of atom-smashing cyclotrons.

The full-time efforts of a large number of agents were required to combat loose talk, rumor and gossip, much of which, as it turned out upon investigation, was innocent speculation. They trailed honest remarks through churches, sermons, columns, editorials, and wayside restaurant conversations, not because the speakers were suspected but because it was necessary to head off conversation if the world's greatest secret was to be kept. An Oak Ridge bus rider once found himself the subject of an investigation after he had innocently asked a companion, while working a cross-word puzzle, the symbol for Uranium.

On one occasion in 1944, a report reached the Intelligence Unit at Oak Ridge that a preacher at Maryville, 35 miles from Oak Ridge, had dangerously skirted the subject of the atom in a sermon. An Agent called upon the pastor. He discovered that the preacher, having read a speculative article on atomic energy in an old magazine, had used it as a sermon topic, stressing the relationship of the Divine Being to the mysteries of the possible new force of the atom. The Agent and the pastor reached a gentleman's agreement that the topic would not be used again.

This incident was not as perplexing as another which occurred.

From Kansas City, Missouri, one day in 1944 came a long distance telephone call to Oak Ridge from a woman seeking to locate a brother working on the project. The operator reported his name not listed and asked the caller if she was sure she was calling the right place.

"Sure," the woman replied, "Oak Ridge is the place where they're smashing those atoms, isn't it?"

That was probably the first time the operator had ever heard the word "atom" mentioned in connection with the project. But she notified Security immediately. Within a few days Agents had found the woman, who said she was merely passing on rumors her brother had written her from Oak Ridge. Agents reprimanded both her and the brother and they agreed to be more circumspect.

One mountaineer was at one time positive he had discovered the project's objective by keeping his ears open and his mouth shut. "I know what they're making," he confided to a friend. "It's *sympathetic* rubber." Such deductions were mingled with such comment as "they're making fourth-term buttons" and "suit-cases for Eleanor."

In 1944, an article appearing in an Atlanta newspaper furrowed the brows of Security Agents. Knoxville was then riding the crest as the hottest war-boom town in the South. The newspaper sent a reporter to check on Clinton Engineer Works.

"It's hush-hush all the way," his article said, "in a manner to make writers of fiction despair of their imagination.

"A weird picture emerges out of the welter of rumor, conjecture, guesses and attempts to put two and two together with the logical result of four.

"There are thousands of workers at the plant, all pledged to unnecessary secrecy for the simple reason they do not know what they are working on. They go about like orderly, well-paid ghosts.

"The 'heart' of the plant is said to be a building surrounded by three barbed-wire enclosures. The wire is charged with electricity. Guards sit atop towers near the barbed-wire walls, rifles ready, prepared to shoot intruders.

"The latter, too, seems to be carrying precaution a bit too far. Because intruders couldn't possibly get near the so-called heart of the plant. Why? Because workers are finger-printed when they go from department to department. It has been reported that one worker was finger-printed 27 times in a single day.

"What goes in? And what comes out? Mystery again. No one has ever seen materials go in or products come out. Yet the plant covers vast acreage, and its edges have the aspect of a construction camp deep in a South American jungle.

"The plant is in operation. Yet workers see no smoke, hear no noises, feel no rumbles, touch no tangible evidence. The place is said to be alive with military intelligence men, checking, guarding, observing, keeping the deep, dark and doubtless deadly secret.

"Rumors run rife, however. Folks conjecture all the way from new gases to new ammunition, to synthetic rubber, to new explosives, to rockets to the moon, to smashing of the atom, to fourth-term buttons for Roosevelt."

The reference to the atom brought the steam of the Intelligence Division to a full head.

A news commentator for the Mutual Broadcasting Company also caused much consternation in Washington and Oak Ridge in August 1944 when he mentioned Columbia University (a key point in the project's research phase), Oak Ridge, atomic energy and new explosives on his broadcast.

An investigation revealed that a radio reporter had obtained a few basic facts from a scientist who had been associated in a minor way with the work at Oak Ridge. With the meager information, the reporter called upon his imagination. The commentator's staff thought the material so improbable, however, that they laid it aside until one night when copy was short. Then it went into the news script merely as a "filler." The reporter who originated the material explained on being interviewed that he was a pacifist and had written a seemingly exaggerated script in the hope it would reach the Germans and terrify them into requesting an armistice. But his exaggeration was too near the truth.

The editors and the radio newsman in these two incidents

obviously were either not aware of or had overlooked the following confidential note which had been sent out by the Office of Censorship in Washington on June 28, 1943, to 20,000 news outlets over the country:

"You are asked not to publish or broadcast any information whatsoever regarding war experiments involving:

"Production or utilization of atom smashing, atomic energy, atomic fission, atomic splitting, or any of their equivalents;

"The use for military purposes of radium or radioactive materials, heavy water, high voltage discharges, equipment, cyclotrons;

"The following elements or any of their compounds: Polonium, uranium, ytterbium, hafnium, protractinium, radium, rhenium, thorium, deuterium."

Many editors, reporters and commentators undoubtedly associated this memorandum with the elaborate goings-on at Clinton Engineer Works, but for the individual without access to such warnings, the project became somewhat of a fantasy. Two occurrences in 1943 and 1944 pointed up the prevailing confusion.

Shortly after barriers and admission gates were established at Elza, Edgemoor, Solway and Oliver Springs entrances on April 1, 1943—they were to remain until March 19, 1949, when the community of Oak Ridge itself was opened to the general American public—a curious old gentleman turned up at Elza. He was an herb doctor. His pockets were as full of roots as country new ground.

He eyed the guard furtively and then approached him.

"I understand," he said to the man barring his way, "that they're building a new White House in there," explaining that he desired to file for a position with the White House staff.

[73]

"Beat it," the guard informed him gruffly. "We don't know what's being built in there."

The answer didn't please the old man at all. "Well," he said most pointedly, "if you don't know what's being built, how do you know it isn't the White House?"

The old gentleman was as confused as one of the many Anderson County farmers who left their workaday world to obtain employment on the "project." This particular employee was a farmer who also peddled eggs to residents of Norris, near Oak Ridge. For a long time, his customers missed him but one day in 1944 he showed up again.

One of his customers asked where he had been.

"I've been working on the project," he confided.

"Were you discharged?" the customer asked.

"No," the farmer replied. "I just quit. And I'll tell you why.

"There they were, all those thousands of people. They were all getting good money, same as I was. And there were more buildings than you can imagine, and a lot of new roads and a lot of other new things, and they were costing a heap of money.

"I thought it all over, thought it over long and hard. And, to tell you the truth, I had in mind that whatever it was the Government was making over there, it would be cheaper if they went out and bought it."

A simple, frugal man of the hills, the scope of the project had overwhelmed him.

Espionage

Two spy cases involving atomic energy stirred the American people in post-war years.

In 1946 a British scientist, Dr. Alan Nunn May, was sentenced to 10 years in prison by a British court for slipping samples of uranium to a Russian agent. He was accused of violating Britain's official secrets act while engaged in atomic research in Canada in 1945, work which also brought him in contact with atomic energy developments in the United States.

On February 1, 1950, another British scientist, Dr. Klaus E. J. Fuchs, was arrested by his Government on charges of passing atomic energy information to the Russians and sentenced to a prison term of 14 years.

Fuchs, member of a British wartime mission to the United States, held an important post at the Los Alamos project, witnessed the test explosion of the first atomic bomb at Alamogordo, New Mexico, in July, 1945, and had access to a great deal of information on atomic matters, according to the charges filed against him.

Fuchs was accused of twice passing atomic secrets to foreign agents—once in the United States in 1945 and again in Britain in 1947. In 1947, Fuchs visited the United States again as a member of a British delegation to participate in discussions of atomic energy information with the United States, Canada and Britain.

May and Fuchs did not work at atomic energy plants at Oak Ridge; their activity carried them elsewhere in the United States.

The Fuchs case led to the arrest by the Federal Bureau of Investigation of several Americans. They included Harry Gold, a Philadelphia research chemist; David Greenglass, a machinist who served at Los Alamos during World War II as an Army technical sergeant; Alfred D. Slack, a chemist who worked during World War II at both Oak Ridge and Kingsport, Tennessee; Julius Rosenberg, a New York engineer, and Morton Sobell, a New York electrical engineer.

[75]

Lie Detector

In post-war years, Security has been strengthened at Oak Ridge through the use of a machine called the Lie Detector, known technically as a Polygraph.

In use by the Atomic Energy Commission at Oak Ridge since 1946 through an arrangement with the firm of Russell Chatham, Inc., the Lie Detector has been used to give more than 20,000 examinations in the interest of national security. Of all persons who successfully passed the Polygraph examinations, none has been denied security clearance, and in no case has a subsequent investigation contradicted the findings of the Polygraph. The Polygraph does not supplant any other means of investigation, but is another tool in a well-rounded security program.

The Lie Detector is used by banks, jewelry stores and other commercial establishments over the country to check employees. In Oak Ridge, the Lie Detector examinations are designed:

To detect in employees motives or intentions inimical to security and not otherwise disclosed;

To detect a change in an employee's motives and intentions from those compatible with the security requirements of the job to those inimical to security of the project;

To prevent and detect any unauthorized disclosures of information;

To prevent and detect any theft or improper handling of classified material or equipment; and

To determine whether an unauthorized disclosure of information was the result of inadvertence or was intentional and to determine whether a report of improper handling of classified material was inadvertent or intentional.

CHAPTER VI

The Wonders of a New World

THE word "amazing," when used in connection with the building of the town of Oak Ridge and the atomic energy plants, loses its luster of description. It has been said that the work was performed on a "time schedule only slightly short of impossible."

Some strange and wondrous facts emerged in the building of the unprecedented facilities. For instance:

Twelve million square feet of blueprints, a number which would furnish a reader with his favorite newspaper for 650 years, if that newspaper averaged 28 pages on weekdays and was three times that size on Sundays, were used by Stone and Webster Engineering Corporation alone in drafting plans and specifications for the electromagnetic plant;

Planning for the gaseous diffusion plant by the Kellex Corporation required 20,000 pages of specifications, 12,000 drawings and 10,000 pages of operating instructions;

Because copper was short and time was more valuable than gold, 14,000 tons of silver having a monetary value of over $500,000,000 was borrowed from the United States Treasury and used for electrical conductors and bus-bars in the electromagnetic plant;

Hundreds of East Tennessee high school girls, with not the

faintest idea of what their jobs were about, were trained by the Tennessee Eastman Corporation to operate complex dials on electronic equipment used to control delicate machinery in the production of uranium-235;

A total of 6,200,000 linear feet, or 1175 miles, of piping—water lines, oil lines for cooling purposes, chemical lines, vacuum piping and nearly nine miles of glass piping—and almost one quarter of a million valves were installed in the electromagnetic plant;

Magnets nearly 100 times as large as any previous magnet ever built and containing thousands of tons of steel were installed; they were scores of feet long, so powerful that their pull on the nails in a pair of shoes sometimes made walking difficult and snatched wrenches from workmen's hands if the tools were loosely held;

Pumping equipment was designed and built capable of producing a vacuum 30,000,000 times that commonly used in standard power plant practice;

Essential tolerances necessary for successful operation of equipment in the gaseous diffusion plant and the uranium chain-reactor at Oak Ridge National Laboratory are much finer than the works in the finest watch;

Leak-proof pumps were developed for use at the gaseous diffusion plant to operate at velocities greater than the speed of sound; the time spent in research, development and design of the pumps alone totaled 250,000 man hours or the equivalent of one engineer working every day for 100 years;

Porous barriers for the concentration of uranium-235 by the gaseous diffusion method were developed which not only contained billions of holes smaller than two-millionths of an inch in diameter but had to be amenable to manufacture in large quantities, measured in acres;

[78]

Cleanliness is so vital in the gaseous diffusion operations that a thumbprint represents contamination; so critical was the problem of producing welded joints to meet tightness and cleanliness specifications that special welding schools had to be held; total welding machines in simultaneous operation were 1200, the total length of welding was: Arc welding, 400 miles, atomic hydrogen welding, 100 miles and airtight welding on enclosures, 100 miles;

A total of 3800 miles of electrical conductors and 825 miles of electrical conduits, involving more than 90,000 separate tests on electrical systems, were built, and half a million valves installed, together with thousands of precision instruments requiring 4,000,000 feet of copper tubing and 3,000,-000 feet of copper wire—all in the gaseous diffusion plant, the largest continuous chemico-physical process in the world;

The operating floor of the gaseous diffusion plant is so vast that technical personnel use bicycles to reach recording instruments stretching for a distance of nearly half a mile,

The 37,562,000 board feet of lumber used in building the electromagnetic plant, if utilized for homes, would equal 7,680 average-sized individual dwelling units, enough for a town of 23,040; the 350,000 cubic yards of concrete used in the gaseous diffusion plant is equal to a solid block of concrete 105 feet high, the size of a city block; the 40,000 tons of structural steel used in the gaseous diffusion plant would build a railroad line 303 miles long; the 5,000,000 bricks used in the same plant would, if placed end to end, stretch 632 miles and the 15,000 tons of sheet steel used in the same plant would provide cooking utensils sufficient for 600,000 average families;

A steam power plant costing $34,000,000 to serve the gaseous diffusion plant, the largest steam plant ever constructed in one operation, has generating equipment with a maximum capacity of 238,000 kilowatts (twice the capacity of Norris Dam of the Tennessee Valley Authority).

These are but a few of the strange statistics on the building of the atomic energy facilities.

The present-day main operating plants at Oak Ridge are the gaseous diffusion plant, the electromagnetic plant and the Oak Ridge National Laboratory, nuclear research center, but another plant which made a vital contribution to the war effort by supplying additional quantities of uranium-235 was the thermal diffusion plant.

On June 26, 1944, the H. K. Ferguson Company of Cleveland, Ohio, was retained as architect-engineer-manager for the building of the thermal diffusion plant after Dr. P. H. Abelson, of the Navy's Research Laboratory at Washington, proved the operability of the thermal diffusion method for obtaining U-235.

Ferguson Company officials estimated that the plant could be designed and built in six months if given top priority. Manhattan District officials instructed that it be finished in four months, later revising the completion schedule to three months. Actually, the plant was in operation in 75 days flat, furnishing feed material to the electromagnetic units. Among material used in the thermal diffusion plant, which cost $10,-500,000, were 50 miles of nickel pipe and 20 miles of iron pipe. Sub-contractors included the Edenfield Electric Company, Turner and Ross Co., Pittsburgh Pipe and Equipment Co., the National Valve Co., the Tri-State Asbestos Co., the Tennessee Roofing Co., the Pacific Pump Co., Westinghouse, Mehring and Hanson Co., and the Grinnel Corporation. The Fercleve Corporation, a subsidiary of the H. K. Ferguson Co., operated the unit.

Having made its contribution to the war effort, the thermal

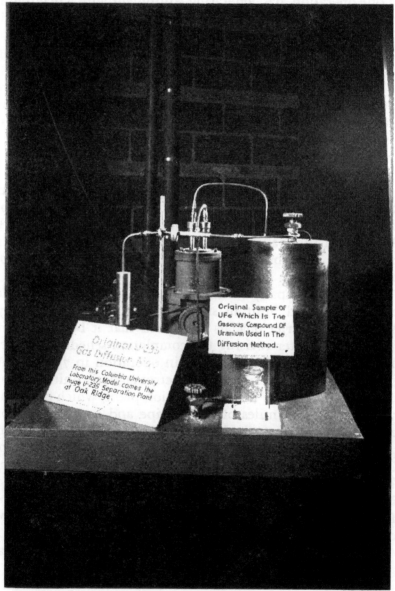

Original Sample Of
UF6 Which Is The
Gaseous Compound Of
Uranium Used In The
Diffusion Method.

Original U-235
Gas Diffusion Model.

From this Columbia University
Laboratory Model comes the
huge U-235 Separation Plant
at Oak Ridge.

This small unit served as a model . . .

. . . for the building of a huge gaseous diffusion plant for obtaining U-235

Atomic Energy Commission

The electromagnetic plant produced the first U-235 on a large scale

The Oak Ridge National Laboratory is of world-wide significance in research

diffusion plant was placed in standby October 18, 1945, and later discontinued.

The Gaseous Diffusion Plant

Design work on the gaseous diffusion plant was started in 1942 by the Kellex Corporation, a unit of the M. W. Kellogg Corporation of New York City, which also handled the supervision of construction and procurement of equipment. Fundamental research for the gaseous diffusion plant was carried on principally at Columbia University under the leadership of Dr. H. C. Urey and Dr. John R. Dunning.

The J. A. Jones Construction Company of Charlotte, North Carolina, built the main plants while Ford, Bacon & Davis, Inc., of New York City designed, constructed and for a time operated a huge auxiliary plant to condition equipment before placement in the process plant. The main gaseous diffusion process building is a huge U-shaped structure, 2,450 feet long and averaging 400 feet in width and 60 feet in height. The total area of the main building is 44 acres. Close by are other huge gaseous diffusion process areas, one of which was completed after World War II ended and two others on which construction was begun in 1949 and 1950 by the Maxon Construction Company, of Dayton, Ohio.

The plant area contains many additional buildings, bringing the total area to around 1,000 acres. Special warehouses contain tens of thousands of different types of spare parts. A special railroad spur branching off the Southern System was built to serve the area. It was operated during the war by J. A. Jones Company; the Southern Railroad now operates it.

Construction of the main process plant was started September 10, 1943, by the Jones Company and the first units for the production of U-235 began operating February 20, 1945. The

[81]

construction forces on this plant reached 25,000 in May, 1945. The entire gaseous diffusion plant, which cost around $500,-000,000, is operated by Carbide and Carbon Chemicals Division, Union Carbide and Carbon Corporation. Carbide and Carbon signed a contract with the Manhattan District effective January 18, 1943, to operate the plant. Carbide furnished technical consultants and service to aid in the design, engineering, construction, and finally to operate, upon its completion, the gaseous diffusion plant. Clark E. Center is general superintendent for Carbide in Oak Ridge.

The Chrysler Corporation, Allis-Chalmers, the American Machine Defense Corporation, the Valley Iron Works, Appleton, Wis.; the F. J. Stokes Machine Co., the Beach-Russ Co., New York; Westinghouse, the National Research Corporation, the Midwest Piping and Supply Co., Houdaille-Hershey Co., the Taylor Instrument Co., General Electric, the Pacific Pump Works and the Wolverine Tube Division of the Calumet and Hecla Consolidated Copper Co. were among the many industrial organizations which contributed to solving the extremely complex problems of design and manufacture. Du Pont, Harshaw Chemical Co., Hooker Electrochemical Co., and Columbia University, Johns Hopkins, Ohio State, Purdue and Princeton also made significant contributions, as did the Mallinckrodt Chemical Works of St. Louis, Mo.

Principal structural sub-contractors under J. A. Jones Co. were D. W. Winkleman, Syracuse, N. Y.; Oman-Creighton Co., Nashville, Tenn.; Wolfe-Michael Construction Co., St. Augustine, Fla.; Bethlehem Steel Co., Bethlehem, Pa.; Virginia Bridge Co., Roanoke, Va.; Lambert Brothers, Knoxville, Tenn.; Birmingham Slag Co., Birmingham, Ala.; Cooney Brothers, Tarrytown, N. Y.; Bryant Electrical Co., High Point, N. C.; Midwest Piping & Supply Co., St. Louis,

Mo.; Poe Piping & Heating Co., Greenville, S. C.; G. G. Ray & Co., Charlotte, N. C.; Kerby-Saunders Inc., New York, N. Y.; and the Ce-Mas-Co Floor Co., Chicago, Ill.

Sixty thousand carloads of materials were used to build this plant alone.

The gaseous diffusion plant provides large-scale separation of the uranium isotope 235 from a chemical compound of uranium by gaseous diffusion through porous barriers—barriers which must contain billions of holes smaller than two-millionths of an inch, withstand a pressure head of 15 pounds per square inch, and can *not* become enlarged or plugged up as a result of corrosion or dust coming from elsewhere in the system.

The process involves several thousand stages in which half of the gas processed in each stage diffuses through the porous barriers as enriched U-235 product and is then sent on to the next higher stage for further concentration. The impoverished half is re-pressured and re-cycled through the next lowest stage. The volume of gas re-cycled is enormous—over 1,000,000 times the volume of the enriched gas. The principal behind the separation of U-235 from natural uranium (U-238) is to convert the solid metal into a gas and make use of the difference in the velocity of the two isotopes in diffusing through the porous barriers. U-235 being the lighter, it has a faster diffusion velocity, so the gas eventually diffused through the barrier is richer in U-235 than the feed gas. After passing through the several thousand stages, an appreciable concentration of U-235 is realized.

Essential tolerances and complexities of the plant were such that many advisors considered the plant impossible to build, and many more felt that even if built it would not work. Indeed, one eminent American physicist requested Manhattan District officials, after $100,000,000 had been spent on construction, to discontinue construction since he felt certain technical problems could not be overcome. The problems were reconciled, however, through the ingenuity of American industry, as were other problems that called for excursion beyond any known method of design and construction.

Industry will eventually reap rich rewards from technical advancements made in the development of this plant.

Industrial applications of techniques developed in the construction of the U-235 plants seem pertinent for the petroleum and chemical and processing industries (improved pumping, possible new methods for gasoline fractionation, new type of heat exchanges and improved automatic control); manufacture of pressure and vacuum vessels (pre-testing vessels for leaks and improved vacuum techniques); high vacuum industries (improved methods for vitamin distillation, new method of maintaining high vacuum and low pressure, low temperature dehydration); industries employing corrosive chemicals (new pump and valve lubricants, new treatment of metal surfaces, and induction operated pumps); refrigeration industry (increased safety in equipment and improved handling of fluorides) and electrical industry (new electronic techniques in high vacuum and improved micro-sensitive instrumentation).

The Electromagnetic Plant

The Stone and Webster Engineering Corporation of Boston designed and constructed the electromagnetic plant in co-operation with technical experts from the Radiation Laboratory of the University of California. The cost was approximately $427,000,000.

Ground was broken for the first plant building on February 1, 1943. The first production building was put into use by the operating company, the Tennessee Eastman Corporation of Kingsport, Tennessee (a subsidiary of Eastman Kodak) in January, 1944. The Tennessee Eastman Corporation, which entered into a contract with the Manhattan District in January, 1943, collaborated with Stone and Webster and University of California personnel to check and perfect equipment designs and furnished technical consultants and service.

Peak employment of construction workers totaled 13,200. The peak of operational personnel was 22,000 in 1945. The plant has a total of 170 buildings with a floor area of 4,500,-000 square feet and covers approximately 500 acres.

Principal construction sub-contractors under Stone and Webster were Harrison Construction Co., Pittsburgh, Pa. and Maryville, Tenn.; Ralph Rogers Company, Nashville, Tenn.; Transit-Mix Concrete Corp., New York, N. Y.; D. W. Winkleman, Syracuse, N. Y.; C. O. Struse & Company, Philadelphia, Pa.; Fluor Corp., San Francisco, Calif.; Bethlehem Steel Co., Bethlehem, Pa.; Watson-Flagg Engineering Co., New York, N. Y.; Hanley & Company, Chicago, Ill.; Rockwood Sprinkler Co., Worcester, Mass.; Tennessee Roofing Company, Knoxville, Tenn.; Bristol Steel & Iron Works, Bristol, Va.-Tenn.; Drainage Contractors, Inc., Detroit, Mich.; and Sullivan, Long & Haggerty, Bessemer, Ala.

On May 5, 1947, the plant operations were taken over by Carbide and Carbon Chemicals Division, Union Carbide and Carbon Corporation, after the Tennessee Eastman Corporation, which received a War Department salute for its splendid work, asked to be relieved of its responsibility because of the end of the wartime emergency.

The electromagnetic plant involved problems of design and construction never before encountered. Since it became

[85]

the first and only plant of its kind in the world, there was no time to construct even a small pilot plant to carry out the methods of separating the uranium atoms (U-235 from U-238) under the electromagnetic process as developed by Dr. E. O. Lawrence of the University of California, who receives the largest share of credit for the scientific development which made possible the remarkable transmutation of a laboratory model into a great industrial plant.

General Electric, Westinghouse, Allis-Chalmers and Link Belt Co., among many others, were the manufacturers and suppliers of equipment.

The theoretical operation of the electromagnetic type of plant involves ionizing uranium particles, accelerating a continuous stream of these particles in a closely-defined path to a speed approaching that of light, bending this stream into semi-circles by means of a powerful magnet field in an almost absolute vacuum, and then catching the particles of U-235 and U-238 in different containers as soon as they become separated. The semi-circular paths of these ionized particles have radii proportional to their momenta. Accordingly, the U-235 is mainly in an arc which has a greater radius than the arc containing U-238.

The electromagnetic plant was flexible—one reason for its selection. Units were built in groups, although most of the controls were separate from each unit. Thus, it was possible to build the plant in steps and start operating the first unit before the second was begun. Design of subsequent units was changed as construction proceeded. So successful was this program that obsolescent equipment was replaced rapidly by new designs, which ultimately resulted in such an improvement in process that it was possible to carry on operations with much less personnel than the peak operating force of 22,000 employees.

The U-235 produced in the electromagnetic plant during the war was so precious that Stone & Webster designed a chemical salvage plant 100 times as large as any test plants that could be used as guides. Every possible grain of U-235 was reclaimed—from work clothes, from water and steel, and even in the air in the plants. And all of the production units had to be controlled through amazing automatic mechanisms operated by personnel of average intelligence (many of the employees were high school girls), who had not the faintest idea what their jobs were about, but who operated dials to produce the material which, when used, liberated a part of the power of the universe.

Oak Ridge National Laboratory

The Oak Ridge National Laboratory was designed and built by E. I. du Pont de Nemours and Company. It was originally a small pilot plant on which the design of the huge plutonium separations units at Hanford Engineer Works in the State of Washington was based.

Construction began February 1, 1943, and operations of the first uranium chain-reactor in the world with a production·potential began at 5 A.M., November 4, 1943. First significant amounts of plutonium for research leading to the building of the Hanford Plant were obtained here.

The plant covers more than 150 acres and cost approximately $13,000,000. Peak construction employment reached 3,247. Peak operating employment in wartime was 1,234; now it is much larger.

From the time of completion until July 1, 1945, the plant was supervised by the Metallurgical Laboratory of the University of Chicago. It was then taken over by the Monsanto Chemical Company of St. Louis, for operation as a nuclear research center. On December 31, 1947, the Atomic Energy Commission announced that Carbide and Carbon Chemicals Division, Union Carbide and Carbon Corporation, would operate the Laboratory beginning March 1, 1948. On February 1, 1948, the Atomic Energy Commission announced that the new name for the facility would be Oak Ridge National Laboratory. It was formerly known as Clinton Laboratories.

This area contains more than 160 buildings. These include three chemistry buildings, a technical laboratory, a pile (uranium chain-reactor) building (a pile of specially designed graphite into which slugs of U-238 are placed), a physics laboratory, a power house, an instrument laboratory, research development shops, a lead shop, a metallurgical building, a medical building (headquarters for the Health Physics group which controls and studies radiation hazards) and several administration buildings and warehouses.

The Laboratory has been enlarged in the past few years in keeping with its designation as a permanent National Laboratory.

The pile at Oak Ridge National Laboratory is the source of production and distribution of radioactive isotopes which are now being widely used for research in medicine, biology, agriculture and industry. Since August 2, 1946, thousands of shipments have gone to various research groups over the country—and abroad. The distribution of these radioisotopes (radioactive isotopes are variations of common elements with the same chemical properties as the stable element but having a different atomic weight and exhibiting the property of radioactivity) is one of the most important peacetime applications in the development of atomic energy.

Other peacetime work being carried on at Oak Ridge National Laboratory in nuclear research include studies in tracer chemistry; biological research; a health physics pro-

[88]

gram which not only controls but continues to study improvements on the control of radiation hazards, and research on uranium chain-reactors looking toward eventual use of atomic energy for commercial power purposes. Increased emphasis has been placed on basic research and chemical process development work at the Laboratory.

The Steam Power Plant

A high-temperature, high-pressure steam power plant lies near the gaseous diffusion process and is an important adjunct in the process carried on in the production of U-235. The plant was designed and developed by the Kellex Corporation and Sargent and Lundy of Chicago and constructed by the J. A. Jones Construction Company at a cost of $34,000,-000.

Sub-contractors included A. S. Schulman Electric Co. of Chicago, William A. Pope Co. of Chicago, The Foundation Co. of New York, the Bethlehem Steel Co., Combustion Engineering Co., Research Corporation, Allis-Chalmers, General Electric and Westinghouse.

It is the largest steam plant ever constructed in one operation and has generating equipment with a capacity of 238,000 kilowatts (twice Norris Dam of the TVA). The paradox of this gigantic steam plant in the heart of the TVA region is explained by plant requirements for variable frequency current and the necessity for minimizing possibilities of interruption of power. The use of the steam plant is in addition to such power as is available from the TVA, the general source of supply for all of the Oak Ridge Area.

The power house contains three boilers, each designed to produce 750,000 pounds per hour of superheated steam at a pressure of 350 p.s.i. and a temperature of 950 degrees and 14 turbo-generators ranging in size from 1500 to 35,000 kilowatts. Adjacent to the power plant there is a main switch house, and an auxiliary switch house, a pump house, a 154 KV switchyard, a service building and the necessary facilities for storage and handling of coal.

Construction of the power plant started June 1, 1943, and initial operation was begun April 13, 1944. It reached its full generating capacity in July, 1945. Until 1950, coal was used solely as fuel; now natural gas from the Texas fields also is utilized under a contract between the Atomic Energy Commission and the East Tennessee Natural Gas Company. The plant is operated by Carbide and Carbon Chemicals Division, Union Carbide and Carbon Corporation.

A Potent Factor

Carbide and Carbon Chemicals Division, Union Carbide and Carbon Corporation, is the most powerful single factor in present-day Oak Ridge operations.

This great industrial organization entered the atomic energy program early in the history of the activity when, effective January 18, 1943, its officers signed a contract with the Manhattan District. The terms called for Carbide to "furnish technical consultants and service to aid in the design, engineering, construction, and finally to operate upon its completion, the gaseous diffusion plant at Oak Ridge." In carrying out this obligation, the people of Carbide made significant contributions to many phases of the facility. Similar work was done by Carbide on the huge steam power plant nearby.

In May of 1947, the Atomic Energy Commission requested Carbide to accept the further assignment of operating the facilities of the Y-12 electromagnetic separation plant. In March, 1948, an additional facility at Oak Ridge was placed in their hands to operate and manage, the Oak Ridge National Laboratory.

Thus the production and research facilities of these three operations are now coordinated under a single management.

The technical know-how and skill of Carbide personnel has brought many improvements, greater production, and notable progress in Oak Ridge operations.

Participation by Union Carbide and Carbon Corporation in the atomic energy program in World War II went far beyond operation of the gaseous plant by one of its units.

The Corporation built and operated plants for the processing of raw materials and assumed management of laboratories formerly under university supervision. The United States Vanadium Corporation and The Linde Air Products Company, both Corporation units, built and operated certain strategic plants; Electro Metallurgical Company, a unit, manufactured special alloys; Bakelite Corporation, a unit, produced important plastic materials; and National Carbon Company, Inc., another unit, developed and manufactured certain critical carbon products.

Billions and Billions

Of the more than $5,000,000,000 that has been spent or appropriated for atomic energy developments in the United States since 1942, a total of over $2,000,000,000 has been for Oak Ridge building and operations.

When the atomic story was released to the world in August 1945, a total of $1,106,393,000 had been spent on Oak Ridge, $382,401,000 on Hanford, and $59,429,000 on Los Alamos.

From 1942 until August 6, 1945, Congressional appropriations for the work were obtained by the War Department through requests which covered up the actual description of the project; in fact, neither the Manhattan District nor its objective was mentioned. A few key Congressional leaders, having been made aware by Secretary of War Stimson of the purpose of the development, guided the "covered-up" requests through Congress. Among them were Senators Alben Barkley, Kenneth D. McKellar of Tennessee and Styles Bridges of New Hampshire, and Representatives Sam Rayburn of Texas and Clarence Cannon of Missouri.

X-10, Y-12 and K-25

The strange combinations of letters and figures as designations for two of the Oak Ridge facilities, X-10 and Y-12, have no particular significance. For the third, K-25, there is a meaning.

The X as applied to Clinton Laboratories (now Oak Ridge National Laboratory) came from the fact that "X" was used by the University of Chicago in its original description of the site. The addition of the "10" had no significance.

The "Y" as applied to the electromagnetic plant was used because it was equally unlikely to have any particular meaning. The "12" was added for no particular reason, except for the purpose of confusion.

The "K" as applied to the gaseous diffusion plant was used because the plant was designed by the Kellex Corporation. Because "25" was used throughout the project to designate U-235, it was added arbitrarily. With the building of other gaseous diffusion plants, K-27, K-29 and K-31 were applied as designations to identify them as separate units.

During the war, the Oak Ridge Area as a whole was known as "Site X," the Hanford area as "Site W" and the Los Alamos area as "Site Y." The letters were utilized simply for cover-up purposes.

[92]

CHAPTER VII

Atomic City Housekeeper

IF, on May 6, 1902, when a young engineer by the name of Henry C. Turner launched the Turner Construction Company in New York City, he had been told that a strange adventure, entirely outside the sphere of normal construction activity, would befall his firm in 1943, he probably would have reacted as did a workman at Oak Ridge when the news of the atomic bomb blast over Hiroshima struck the world. Said the workman: "I am doggoned."

But the Turner Construction Company, which had for its first contract in 1902 the building of a small reinforced concrete vault for the Thrift, a savings institution in Brooklyn, and since that time has produced close to a billion dollars in construction in this country and abroad, had, like many other firms in the United States, a date with destiny.

In 1943, the Army, its hands full fighting a war and creating an atomic bomb, had neither the time nor the inclination to operate and manage a community which was beginning to rise in the hills of East Tennessee. When the need developed for a private organization to operate Oak Ridge, the Manhattan District, which had been impressed by the job the company had done for the Army's Corps of Engineers at an Air Depot at Rome, New York, turned to Turner.

Thus was born the Roane-Anderson Company, which entered the Oak Ridge scene September 23, 1943, under a contract with the Manhattan District as landlord and housekeeper of the Atomic City. Turner Construction Company, a firm headed by Quakers, organized Roane-Anderson, named for the two counties in which the project is situated, as a management firm under Tennessee charter to relieve the Army of the details of running a community. The company established its project office October 7, 1943.

A great and imposing array of America's construction and industrial organizations, together with the country's leading universities, helped bring about the miracle of Oak Ridge.

But the organization which affected the day-by-day life of each Oak Ridge resident, especially the housewife, most directly, was Roane-Anderson.

The average housewife might not have known the facts on plant construction and operations, but she was acutely aware of Roane-Anderson. For Roane-Anderson collected the rent, delivered the coal, picked up the garbage, fixed the faucet, replaced the fuse, repaired the loose plank, replaced the window pane, supplied the maid, and carried out myriad other duties in catering to the needs of 75,000 persons in times of great secrecy and security.

Roane-Anderson did not build Oak Ridge. But as rapidly as facilities—houses, stores, utility and water systems, dormitories—were completed in 1943, 1944 and 1945, Roane-Anderson assumed the responsibility of managing, operating and maintaining the "secret city." Its assignment was to keep the residents in a reasonably satisfied state of mind; the whims of a tenant thus had high priority. In time, an Army officer attached to the Manhattan District described Roane-Anderson as "the best whipping boy the Army ever had."

Turner officials themselves declare that "the management

of the city of Oak Ridge didn't come any closer to being a construction job than Hirohito came to riding his white horse through the streets of Washington," but it was, however, a vast and complicated undertaking.

An idea of the scope and variety of functions Roane-Anderson carried out in meeting the unique challenge of Oak Ridge can be gleaned from the fact that:

It organized and operated a transportation system of over 800 busses, the ninth largest in the United States during the war years, covering over 2,400,000 miles a month and carrying an average of 120,000 passengers per day during peak operations;

It supervised the operations of 17 cafeterias and food-eating establishments which served a million and a quarter meals a month, or 40,000 a day;

It operated a cold storage plant which handled 1,200,000 pounds of perishable merchandise a month, of which 75 per cent was meat;

It ran a farm on the Oak Ridge Area which had a herd of 3,000 cattle and also operated a chicken ranch to supply meat during the wartime shortages;

It operated a 35-mile railroad line with five locomotives and a crew of 105 men in keeping 3,000 cars of construction materials and equipment rolling to the plants each month (officials of the Louisville and Nashville Railroad, which brought the cars to the area boundary and delivered it to Roane-Anderson crews, were puzzled by the fact that they saw thousands of loaded cars going in but only empties coming out);

It operated a million-dollar-a-year laundry business, handling 2,500 dry cleaning and 9,000 laundry customers a week in addition to 100,000 pounds of flatwork;

It supervised the assignment of more than 500 housemaids

and other common labor to the homes of residents and commercial establishments, and for a time employed a group of hostesses to assist newcomers in becoming oriented;

It served as landlord for over 35,000 housing units which were in the city of Oak Ridge proper at peak operations, ran a hotel which had as its guests many of America's most distinguished men, purchased the supplies for a 337-bed hospital, and paid the salaries of hundreds of patrolmen, policemen and firemen;

It carried out what was undoubtedly one of the most comprehensive "hotel" management jobs in history—the sheltering of 15,000 dormitory residents, almost equally divided between men and women;

It sought out and brought in nearly 200 private business establishments and service organizations—department stores, drug stores, shoe shops, barber shops, grocery stores, a bank, dry cleaners—to serve Oak Ridge residents;

It maintained the nearly 300 miles of roads of a new community which grew to such size that if you start from the Empire State Building in New York as one end of the city of Oak Ridge, the other end will be Passaic, New Jersey; and

It employed 10,000 persons at peak operations to carry out its varied duties.

In addition, it maintained schools and other public buildings, operated directly such public services as water supply, the electric system, central steam plants and sewage and garbage disposal and supervised activities which in any other city would be performed by all the various individual property owners, real estate operators and commercial operators.

As in all landlord-tenant relationships, there were at times discordant notes. A telephone conversation in 1944 illustrates the attitude of some.

One afternoon the wife of a Carbide and Carbon official

[96]

telephoned the Roane-Anderson project manager's wife to invite her to a party. The operator informed her the telephone was unlisted.

"But," the wife of the Carbide official said, "I'm not calling on business and I am sure they will not object. Besides, I'm one of their friends."

The operator was courteous but firm.

"Madam," she replied, "for your information, Roane-Anderson has no friends."

The telephone operator's analysis of the situation was no doubt extreme, yet the mushrooming of Oak Ridge during the war years presented complex problems of management. But at the war's end, high praise went to Roane-Anderson from the Manhattan District for the company's contributions to the atomic project.

In the past few years, Roane-Anderson relinquished many of its functions in Oak Ridge, although it retained operational responsibility. Many of its operations were turned over to private enterprise during the transition of the community from a war-born Government-dominated city into one having more of the aspects of the normal American community of the same size.

For example, the railroad into the city is now an L & N-operated line, the restaurants and cafeterias are under private management, the tenants have certain maintenance responsibilities, the 200-bus transportation system is operated by a private firm, American Industrial Transport, Inc., which took over in February 1945; laundry and dry cleaning is accomplished by private firms, and the principal hotel is under private management of Alexander Hotels, Inc. Many other similar steps directed toward normalcy have been taken.

Roane-Anderson Company has done its complex management job under a contract prescribing a certain fixed monthly fee for the company, which turns over all funds collected to the Government.

The contributions of Turner Construction Company in World War II were not confined to running the community of Oak Ridge through its subsidiary, the Roane-Anderson Company. The firm was a part of the far-flung construction forces which built vital Naval bases in the Pacific. It also constructed many important plants for private companies and branches of the Armed Forces. Among such contracts were plants for the Atlas Powder Company, Atlas Point, Del.; Electro Metallurgical Company (a subsidiary of Union Carbide and Carbon Corporation) Ashtabula, Ohio; Hercules Powder Company, Belvidere, New Jersey; General Electric Company, Everett and Lynn, Mass.; Higgins Aircraft, Inc., New Orleans; Naval Shipyard, Brooklyn; Pratt & Whitney Aircraft Corporation, Kansas City, Mo.; the Glenn L. Martin Company, Middle River, Maryland; the United States Rubber Company, Chicopee Falls, Mass., and the U. S. Air Base at Rome, N. Y.

In the latter part of 1943, an elderly woman approached the Roane-Anderson employment office in Oak Ridge seeking a job. She called it the "Rogue and Angerson Company."

Asked what she could do and why she wanted to work for Roane-Anderson, she replied that she was interested in "getting that money time and time again." Slightly confused, she had heard that good pay was available through "time and overtime."

"Time and time again," the Turner Construction Company has made significant contributions in helping to build up the nation's industrial capacity.

As one of its projects during World War I, Turner built the War and Navy Office Buildings in Washington under a contract approved by the then Secretary of the Navy, Franklin D. Roosevelt; in World War II it managed a "secret city" as part of a $2,000,000,000 "calculated risk" approved by President Franklin D. Roosevelt.

Turner Construction Company, in two World Wars, has built and managed for the arsenal of democracy.

CHAPTER VIII

A Story Is Born

IN April 1945, events moved swiftly in the saga of the A-Bomb. A few days after his inauguration following the death of President Roosevelt on April 12 at Warm Springs, Ga., President Truman, who as a United States Senator and Vice-President had been aware of the project and its implications, was given a complete, up-to-date picture of the atomic program at a White House conference by Secretary of War Stimson and General Groves.

Simultaneously, the 509th Composite Group, 313th Bomb Wing, 20th Air Force, was completing its tests on runs with dummy bombs at its secret base at Wendover, Utah, where training had started in the Fall of 1944, and was preparing to leave for the Pacific.

On Tinian, in the Pacific, from which point the "Enola Gay," a B-29 with Col. Paul W. Tibbetts, Jr., of Miami, Fla., at the controls was to take off on August 5 with its nose turned toward Hiroshima, construction was well under way on the Atomic Base under the direction of Col. Elmer E. Kirkpatrick, who later was to serve as commanding officer at Oak Ridge.

And in a ceremony in April at the Knoxville, Tenn., Airport, a B-25 bomber, purchased with $250,000 contributed by workers in Oak Ridge's K-25 Area, was christened "Sunday Punch" before being flown to combat in the Pacific.

Earlier in April, the first step was taken in making arrangements for developing the background information for the newspaper releases which were to fall on a startled world in August.

At that time, General Groves and Col. W. A. Consodine, press relations advisor to the General, requested Jack H. Lockhart, assistant to Byron Price, Director of the Office of Censorship in Washington, which collaborated closely with the Manhattan District during the war in maintaining the secrecy of the atomic venture, to take the War Department assignment of writing the official story of the A-Bomb. Because of other commitments, Mr. Lockhart, now Assistant General Editorial Manager for the Scripps-Howard Newspapers, declined, but recommended William L. Laurence, eminent science writer for the *New York Times* and a Pulitzer Prize winner in 1937, whose familiarity with the possibilities of the atom had been revealed as far back as May 5, 1940, in a *Times* article. Mr. Laurence also had written an article for the *Saturday Evening Post*, September 7, 1940, entitled "The Atom Gives Up," a stirring look into the future.

When actual work on the atomic project started in 1942, the *Post* was requested by Intelligence officials to take the September 7, 1940, issue out of circulation. Similar requests were made of libraries over the country and all persons inquiring for that particular issue were investigated.

Mr. Laurence, who was to be the only newspaperman present at the Alamogordo test explosion in New Mexico July 16, and later an observer of the Nagasaki blast, reported for duty with the Manhattan District in Washington on May 8 after General Groves and Colonel Consodine had discussed his availability with Edwin L. James, managing editor of the

Times, in a New York conference. To Mr. Laurence, it was "one of the biggest stories of all time."

The job of putting the official War Department story of the atom on paper began May 10 in Oak Ridge when Mr. Laurence reported to Col. K. D. Nichols, District Engineer for the Manhattan District. On the previous day, Colonel Nichols, in a statement concerning the German surrender May 6, had informed Oak Ridge employees that "the operations at Clinton Engineer Works will continue in full force until complete victory over Japan." At that time, the total strength of the Japanese Army was estimated at 5,000,000.

Mr. Laurence began his quest for official facts with visits into the huge, restricted Oak Ridge plants. The quest then took him to Los Alamos, Hanford, Berkeley, California; Chicago and New York, and back again to Oak Ridge, where he wrote the majority of his stories in a special office set aside for him in District headquarters. In all, Mr. Laurence traveled over 50,000 miles on his A-bomb itinerary, a tour climaxed as he rode in an escort plane when the second bomb fell over Nagasaki August 9.

By July, details had been synchronized for the handling and distribution of the atom story. In Oak Ridge, Lt. George O. Robinson, Jr., formerly of the *Memphis Commercial Appeal,* and Public Relations Officer for the Manhattan District, set up a restricted reproduction room, where 12 hand-picked members of the Woman's Army Corps Detachment at Oak Ridge cut the stencils and produced thousands upon thousands of mimeographed sheets which were to go to the public after the first use in world history of an atomic bomb in warfare.

Meanwhile, in Washington, Colonel Consodine was enlarging his staff in preparation for the big break. Assigned to him there were Lt. Col. Clyde Mathews, formerly with the Jackson

(Mississippi) *Daily News;* Major John F. Moynahan, formerly with the Newark, N. J., *News* and Capt. Kilburn R. Brown, former magazine writer. At Hanford Engineer Works, Lt. Milton R. Cydell, formerly with the Seattle *Times,* was assisting in the preparations. All aided in writing and developing the first releases.

Shortly after the arrival of Mr. Laurence in Oak Ridge, James E. Westcott, staff photographer for the Public Relations Office, was assigned to take aerial and ground pictures throughout the Area. Westcott made over 200 exposures from a chartered plane. Assisted by the late H. B. Smith, another official Oak Ridge photographer, Westcott's office processed over 5,000 prints of the 33 pictures—Oak Ridge and Hanford scenes and pictures of Manhattan District officials—which had been selected for official distribution after clearance by Manhattan District Security and War Department officials. Legends for the photographs were prepared in Oak Ridge and the pictures assembled there. On July 18, two days after the Alamogordo test blast, Westcott, entirely unaware of the significance of his mission, made the last of the official pictures, photographs of General Groves and his assistant, General Thomas F. Farrell, in Washington.

A few days before the end of July, all was in readiness. Thousands of mimeographed pages of 14 separate press releases had been prepared and together with hundreds of sets of the 33 official pictures had been flown to Hanford, Los Alamos and Washington in an Air Force plane under heavy guard to be held for distribution.

Simultaneously, the last shipment of U-235 for the Hiroshima bomb left Oak Ridge July 25, arriving at Tinian July 27. And on July 25, a directive was prepared for the signature of General Thomas T. Handy, acting in the absence of General George C. Marshall, who was in Potsdam, authorizing the 509th Bomb Group to attack Hiroshima, or as alternates Kokura, Niigata or Nagasaki.

With the completion July 27 of the reproduction of the first press releases and pictures, Intelligence and Security agents were dispatched from Oak Ridge, armed with the official handouts, to key Southern cities—among them Nashville, Memphis, Chattanooga, Atlanta and Birmingham—to await word of the dropping of the bomb. Releases for the Knoxville newspapers and radio stations were held in readiness at Oak Ridge and arrangements made to service other news media directly from Oak Ridge.   — ...—

The 14 written releases embraced:

1. Account of the July 16, 1945, New Mexico test.

2. General news story based on President Truman's announcement of the use of the atomic bomb.

3. Story of the discovery of uranium fission.

4. Atomic energy, derivation and theory.

5. The gaseous diffusion plant at Oak Ridge.

6. The electromagnetic plant at Oak Ridge.

7. The thermal diffusion plant at Oak Ridge.

8. The town of Oak Ridge.

9. Hanford Engineer Works.

10. Labor aspects, including recruitment, of the project.

11. Biographical, General Leslie R. Groves, commanding general of the Manhattan Project.

12. Biographical, General Thomas F. Farrell, Assistant to General Groves.

13. Biographical, Colonel Kenneth D. Nichols, District Engineer, Manhattan District.

14. Biographical, Colonel Franklin T. Matthias, officer in charge of the Hanford Engineer Works.

The basic work completed, key officials of the Manhattan District and the War Department stood at alert for a week as

Oak Ridge captured the world's headlines beginning in August, 1945

J. E. Westcott

the S-1, the code name for the new weapon, neared its rendezvous with destiny.

As of August 5, President Truman was somewhere in the Atlantic aboard the USS Augusta on his way home from the Potsdam Conference. (On January 27, 1947, in a letter to Dr. Karl Compton of the Massachusetts Institute of Technology, President Truman said he had personally made the decision to use the bomb.) In Washington, where the War Department Public Relations Department had been advised by the Manhattan District to be prepared for the big story, Colonel Consodine, Colonel Mathews and Captain Brown, together with Robert Ebbinger and Edmund Durkin, were standing by and Lt. Robinson at Oak Ridge and Lt. Cydell at Hanford were awaiting the go-ahead signal.

Major Moynahan was with General Farrell on Tinian awaiting word from the crew of the "Enola Gay," already winging its way toward Hiroshima, with Admiral William S. Parsons, Naval ordnance expert, who was as responsible as any military man for the bomb's development, as the "weaponeer" who armed the new machine of destruction. After the Hiroshima drop, Major Moynahan and General Farrell wrote leaflets which were scattered over Japan cautioning surrender lest still more cataclysmic events befall the homeland.

On Monday, August 6, Washington was quiet. The firstcomers to the White House press room had no way of knowing that the dawn of the atomic age was about to burst upon them. The President was still away. At 10 A.M., Eben Ayres, assistant press secretary, informed the newsmen that "there might be something later."

At 11 A.M., the more than two score newsmen who were by that time present were met by Maj. Gen. Alexander D. Surles, War Department Public Relations Officer. Before him was a press release, the official announcement of the President, who had radioed his approval from the USS Augusta. Additional press releases were on hand at the Pentagon Building.

Mr. Ayres read the first paragraph, handed out copies. In a few seconds, press wires throughout the world were humming with the story of the atomic bomb; little else was to occupy the attention of the press for some time. On August 9, the second atomic bomb fell on Nagasaki, delivered by a B-29, the "Bockscar," piloted by Major Charles W. Sweeney, of North Quincy, Mass. Russia entered the war against Japan and surrender was near. November 1 was the date which had been officially set for the invasion of the Japanese homeland by American troops. The estimates of the expected American casualties varied, but they ran as high as a million. Because of the atomic bombs, thousands of American boys whose fate would have been otherwise were safe and no longer in danger.

President Truman's dramatic August 6 announcement . . .

> "Sixteen hours ago an American airplane dropped one bomb on Hiroshima. That bomb had more power than 20,000 tons of TNT. It is an atomic bomb . . . a harnessing of the basic power of the universe."

. . . fell on Oak Ridge with a strange impact.

Persons with access to radios were the first to hear. The men in the plants who had helped to produce the first U-235 to be used in warfare received the news when their wives telephoned to tell them of what Oak Ridge had done.

"I felt," explained one employee, "that when my wife said some mention had been made of the atom, she might be splitting up herself."

For several hours, there was a strange stirring in the community, and talk was hushed—some men instinctively whispered "sh-sh-shush"—as if further details were necessary to make people believe. By midafternoon, however, the shock had worn off. Great rejoicing coursed through the business section, the plants and the homes. Oak Ridgers were so anxious for information that "extras" of Knoxville newspapers sold for $1.00 a copy; one circulation man disposed of 1600 papers in 35 minutes. In New York, the wife of an official of Union Carbide and Carbon Corporation greeted her husband at the door of their home on his return from work: "On the radio today I heard about a place called Oak Ridge, Tennessee. Is that Shangri-La?" Some Carbide people had used "Shangri-La" as a code word for Oak Ridge during the hush-hush days.

Oak Ridgers, at last, could talk. Certain words once recorded "secret" were no longer secret. After many months of toil and stress and "buttoned" lips, they could mention the atom and they rolled out the word with great relish and exhilaration. The workers who had talked about everything but their work—"your wife, the school system, Heaven and hell, even the price of babies"—could now discuss the project.

With the dropping of the bomb, Oak Ridge became the focal point for scores of newspaper, radio, newsreel and magazine representatives. From special press rooms set up in a dormitory, Casper Hall, by the Public Relations Office, 200,-000 words of copy were moved the first four days alone. Representatives of the press and radio who have visited Oak Ridge since that time and continue to visit one of the world's most unusual cities are numbered in the hundreds.

The revelation of the bomb was an exhilarating experience for Oak Ridgers. They had said "no" for so long that they found it difficult to say "yes." Sworn to silence, the force of habit had become second nature. Never before had so many people said so little about so much.

They confined their discussions to information officially released and to the official statements, one of which, from Colonel Nichols, eased the fears of Tennesseans that there might be danger of an atomic explosion in Oak Ridge itself. "The safety records achieved are superior; many industries have far greater hazards," the jittery were advised.

The exuberance of the occasion reached far and wide.

On the morning of August 7, when the Southern Railroad's train from Washington rolled into the Knoxville station, a Pullman porter announced to his passengers:

"You are now entering Knoxville, the gateway to Oak Ridge."

Today, Oak Ridge is the gateway to a bright, new world in constructive uses of atomic energy.

Roosevelt and Oak Ridge

President Roosevelt did not get to see Oak Ridge.

On April 10, 1945, however, two days before the President's death, Secretary of War Henry L. Stimson made a tour of the plants. Special ramps were built in one of the plant areas to accommodate Mr. Stimson; because of the ramps, word spread that Mr. Roosevelt was en route to Oak Ridge. Instead, he was at Warm Springs.

Mr. Roosevelt, it is recorded, was given his last information on the status of the A-Bomb program on March 25, 1945, by Secretary Stimson at the White House.

Significant Dates

Three significant events occurred during World War II on the Sixth of the Month.

On June 6, 1944, General Dwight Eisenhower's forces stormed the French Coast on D-Day; on May 6, 1945, the

Germans surrendered and on August 6, 1945, the first atomic bomb used in warfare was dropped on Hiroshima, Japan.

Oak Ridge also apparently has an affinity for the month of August. In addition to the first atomic bomb being dropped August 6, 1945, the Manhattan District was established August 13, 1942; the bill establishing a Civilian Commission to take over the atomic project from the War Department was signed into law by President Truman on August 1, 1946, and the first radioactive materials for purposes of peacetime research in biology, medicine, agriculture and industry were shipped from the atomic "furnace" at Oak Ridge National Laboratory on August 2, 1946.

CHAPTER IX

Into Civilian Hands

THE miracle of the atomic bomb was wrought by the most unusual combination of fundamental scientific research, of technical development both in and out of the laboratory, of mass production and of management and labor ever achieved in American history. The results were obtained through the most zealous, active and self-sacrificing cooperation, by hundreds of thousands of Americans in numerous different walks of life, that this country has ever known, all carried out in absolute secrecy and seclusion.

In a stragetic race against time, a generation of national effort was, in effect, compressed into a period of barely five years.

When the hostilities ended, the Manhattan Project, at a cost of nearly $2,000,000,000 had developed two atomic bombs which helped hasten the fall of Japan. The military objective of the project had been reached. In the American tradition, the time had come for transfer of the vast enterprise into civilian hands.

The transfer, however, was slow in evolving. The Congress, with the national and international implications of the new force of atomic energy weighing heavily upon it, was immediately indecisive as how best to set up a civilian organization which would assume responsibility for utilizing atomic energy for improving the public welfare and promoting world peace. The powers to be granted such a civilian organization and the extent to which the military would participate in its program brought on lengthy debate.

After long examination by a special Committee of the Congress established in the Fall of 1945 to examine the nature and implications of atomic energy and to frame the basic law under which atomic energy would be controlled and developed, an Act was finally written and agreed upon in July 1946.

On August 1, 1946, almost a year to the day after the first atomic bomb fell on Hiroshima, President Truman signed into law the Atomic Energy Act of 1946, better known as the McMahon Act, named for Senator Brien McMahon of Connecticut, who introduced the legislation as first chairman of the Senate-House Joint Committee on Atomic Energy.

On October 28, 1946, President Truman announced the members of the five-man Atomic Energy Commission as provided in the Atomic Energy Act. As chairman, he designated David E. Lilienthal, former chairman of the Tennessee Valley Authority, who also had served as chairman of a Board of Consultants to the Secretary of State's Committee on Atomic Energy. The Board prepared a report which was used in the development of the American proposals for atomic energy control to the United Nations' Atomic Energy Commission, commonly referred to as the Acheson-Lilienthal Report. (Dean Acheson, later to be named Secretary of State by President Truman.)

Others members named were Dr. Robert F. Bacher, a leading U. S. physicist who participated in the development of the atomic bomb; Sumner T. Pike, business executive and former member of the Securities Exchange Commission;

Lewis L. Strauss, New York business leader and a Rear Admiral in World War II; and William W. Waymack, editor of the Des Moines *Register-Tribune*.

On January 1, 1947, the Manhattan Project, which owned or leased more land than all of Rhode Island and had in its possession more material and equipment than could be stored in the Empire State Building, was officially transferred to the Atomic Energy Commission. But the Commission itself was not to be formally cleared to take office until April 9, 1947, when, after lengthy and spirited hearings before the Senate Committee on Atomic Energy on the qualifications of Mr. Lilienthal and the other nominees, the Senate confirmed them by a vote of 50 to 31.

From the hearings emerged what is generally regarded as one of the most dramatic and forceful enunciations of its kind on record of an individual's conception of the meaning of democracy and the American way of life. The speaker was Mr. Lilienthal. Questioned on his political views and philosophy, Mr. Lilienthal, in an oral statement delivered extemporaneously under the pressure of a public hearing, said:

"I believe in—and I conceive the Constitution of the United States to rest upon, as does religion—the fundamental proposition of the integrity of the individual; and that all government and all private institutions must be designed to promote and protect and defend the integrity and the dignity of the individual.

"Any form of government, therefore, and any other institutions, which make men means rather than ends, which exalt the state or any other institutions above the importance of men, which place arbitrary power over men as a fundamental tenet of government, or any other institutions, are contrary to this conception; and therefore I am deeply opposed to them.

The Atomic Energy Act was signed Aug. 1, 1946, by President Truman. Looking on were (left to right) U. S. Senators Connally, Millikin, Johnson, Hart, McMahon, whose name the bill carries; Austin and Russell

On Jan. 1, 1947, the War Department (represented by Gen. L. R. Groves) turned over the Manhattan Project to a civilian Commission (represented by David E. Lilienthal, first Commission chairman)

"The Communistic philosophy, as well as the Communistic form of government, fall within this category, for its fundamental tenet is quite to the contrary. The fundamental tenet of Communism is that the state is an end in itself, and that therefore the powers which the state exercises over the individual are without any ethical standards to limit them. That I deeply disbelieve.

"It is important to believe those things which provide a satisfactory and effective alternative. Democracy is that satisfying affirmative alternative.

"One of the tenets of democracy that has grown out of this central core of a belief that the individual comes first, that all men are the children of God and their personalities are therefore sacred, carried with it a great belief in civil liberties and their protection and a repugnance to anyone who would steal from a human being that which is most precious to him, his good name; either by imputing things to him, by innuendo, or by insinuation.

"I deeply believe in the capacity of democracy to surmount any trials that may lie ahead provided only we practice it in our daily lives."

In taking over the atomic energy project, the Atomic Energy Commission found that the uncertainty at the end of the war and the delay in the development of a clear, concise directive on future policies for the great, new enterprise had dulled the momentum of the Manhattan Project. The personnel problem as it related to key scientific workers, many of whom desired to return to the universities and the classrooms, was especially acute. The massive production and research facilities had been unable to go forward with the full speed necessary to maintain this country's pre-eminence in the field of atomic energy. The War Department, acting as custodian of the properties until proper civilian control could be established, had been given no decisive order enabling it to develop a long-range program.

Thus it came about that when the Atomic Energy Commission assumed jurisdiction in 1947, it was faced with a tremendous task. But once confirmed by the Senate, the Commission advanced rapidly in its reorganization of the project and in formulating long-range plans designed to obtain the maximum benefits from atomic energy. By August 15, 1947, when the Manhattan District was dissolved by the War Department—almost five years to the day since it had been established—the Atomic Energy Commission was making swift progress in the development of sound programs for the peaceful application of atomic energy.

Today, the Commission, which under the Atomic Energy Act maintains its headquarters in Washington, guides the destinies of a great enterprise which has made and continues to make rapid advances in the atomic energy field.

Under its jurisdiction come a General Manager with offices in Washington and separate offices of operations in different parts of the country. A Manager is in charge of each Office of Operation. The offices are Oak Ridge, site of the production facilities for uranium-235 and the Oak Ridge National Laboratory, nuclear research center; New York, which has as its primary function the preparation of feed materials for the plants, and also supervises the contract for the operations of Brookhaven National Laboratory on Long Island; Hanford, site of the production facilities for plutonium; Los Alamos, the center of weapons research, development and production; and Chicago, which administers the contract for the Argonne National Laboratory, the third of the Commission's national laboratories. In addition, Chicago administers the contracts for research activities at Ames, Iowa; Berkeley, California; Pittsburgh, Pa., and Schenectady,

N. Y., and Oak Ridge administers the contracts for research activities at Miamisburg, Ohio, and Marion, Ohio, in facilities operated for the Commission by the Monsanto Chemical Company.

The Atomic Energy Commission's program is basically concerned with the production of fissionable and special materials, military applications of atomic energy, the development of nuclear reactors, research in the fields of biology and medicine and research in the physical sciences. Nearly 1300 plants, laboratories, research centers and participating institutions are engaged on this work under contract with the AEC.

Production activity ranges from the exploration for uranium ores to the refinement of plutonium. The discovery, development, mining and extraction of uranium from ores requires extensive geological exploration and the stimulation of activity by private prospectors and mining companies. A great deal of this activity centers in the area of the Colorado Plateau. In addition, large quantities of uranium ore are procured from the Belgian Congo and from Canada.

The New York Operations Office supervises the contracts for converting the uranium concentrates into two principal types of feed material for the production plants. One is pure uranium metal required for the giant nuclear reactors at Hanford. The other is uranium in the form of gaseous uranium hexafluoride which is the feed material for the Oak Ridge production plant. Purity standards normally found only in pharmaceutical materials are demanded.

The final product at Oak Ridge is Uranium-235, an ingredient essential either for atomic bombs or for special nuclear reactors useful for a variety of purposes and ultimately, it is hoped, for the production of power for commercial purposes. The product at Hanford is plutonium, a man-made element produced by the transmutation of Uranium-238 atoms into plutonium, an ingredient also essential for atomic bombs.

Los Alamos is the headquarters for the research, development, production and tests of atomic weapons, which are complicated pieces of machinery and involve many other components beyond U-235 and plutonium and many activities beyond these conducted at Los Alamos.

Many laboratories over the United States are involved in the development of nuclear reactors, the first major atomic energy machine developed other than the bomb. In this field lie the real hopes for the future peaceful applications of atomic energy in the form of power and the production of new and useful materials. A few years may pass before reactors will be built which provide useful energy for the generation of electricity—power for factories, lights for homes and propulsion for ships and planes—but the problem has been attacked with vigor by the Commission in the establishment in Idaho of a national nuclear reactor testing station comprising around 400,000 acres. Users of this field facility, a proving ground for the national reactor development program, include the Oak Ridge National Laboratory, the Brookhaven National Laboratory, the Los Alamos Scientific Laboratory, the General Electric Company and the Westinghouse Electric Corporation.

In the field of biology and medicine, the Commission's three national Laboratories, together with many schools and universities working under contracts administered by the Atomic Energy Commission, are pursuing research programs having far-reaching implications. The effects of radiation upon living organisms and abnormal cell reproduction (cancer) and the effects of radiation on heredity are being studied. At Oak Ridge National Laboratory, a program involving 100,000 mice is being carried out to determine radiation effects and the mutation rate caused by measured doses of radiation. Fruit flies, fish, corn and bread mold also are being utilized in radiation studies in other programs over the country. The field of biology and medicine also includes programs for protection of personnel engaged in atomic energy work, how best to control radioactive wastes and extensive courses in the training of people in the new arts developing from atomic energy.

In the Commission's program for research in the physical sciences, the National Laboratories at Oak Ridge, at Brookhaven and at Argonne are playing leading roles. Other major laboratories also are making vital contributions toward enlarging the still meager understanding of the mysteries of the atomic nucleus.

Other important phases of the atomic energy program are being carried on at Oak Ridge by the Oak Ridge Institute of Nuclear Studies, Inc., the University of Tennessee, the Research Project for the Application of Nuclear Energy to the Propulsion of Aircraft and the Massachusetts Institute of Technology.

The Oak Ridge Institute is composed of 26 member universities in the South and Southwest and chartered under Tennessee law. General purposes are (1) to stimulate cooperation between the Government and participating universities in undertaking fundamental research in atomic energy (2) to foster increased improved programs of graduate studies and education in nuclear energy in southern educational institutions and (3) to aid in the unique facilities in Oak Ridge for graduate research and instruction. The Institute also conducts courses for scientific and medical personnel from over the country in the handling of radioactive isotopes and carries forward a long-range study of the treatment of malignant diseases using radioactive materials.

Member universities are Alabama Polytechnic Institute, University of Arkansas, Catholic University of America, Duke University, University of Florida, Emory University, Georgia Institute of Technology, University of Georgia, Louisiana State University, Mississippi State College, North Carolina State College, Rice Institute, Tulane University, University of Alabama, University of Kentucky, University of Louisville, University of Mississippi, University of North Carolina, University of Oklahoma, University of South Carolina, University of Tennessee, University of Texas, University of Virginia, Vanderbilt University, Texas A & M College and Virginia Polytechnic Institute.

On a 3,000-acre tract in the Oak Ridge Area, a program of general agricultural research has been developed by the Experiment Station of the University of Tennessee under contract with the Atomic Energy Commission. The program includes both animal and horticultural work. Studies have involved a herd of cattle which suffered radiation damages in the atomic bomb test explosion at Alamogordo, New Mexico, July 16, 1945. Various experiments on animals and crops are being made using radioactive materials as laboratory tools of investigation.

The Research Project for the Application of Nuclear Energy to the Propulsion of Aircraft (NEPA) is a combined operation of nine selected companies from the aircraft industry, operating as members, and the NEPA Central Group, headed by the Fairchild Engine and Airplane Corporation, prime contractor under a contract jointly sponsored by the Department of the Air Force and the Navy Bureau of Aeronautics. The Atomic Energy Commission and the National Advisory Committee for Aeronautics are cooperating. The NEPA project comprises investigation of the application of nuclear energy to all possible systems of propulsion of aircraft. It includes studies of closed cycle turbines, open cycle turbines, turbo-jets, ram jets and rocket devices.

A practice school is conducted in Oak Ridge by the Massachusetts Institute of Technology for certain students in its School of Engineering.

At Argonne National Laboratory near Chicago, 31 Midwestern universities and organizations are participating in another far-reaching research program. They are Battelle Memorial Institute, Carnegie Institute of Technology, Illinois Institute of Technology, Indiana University, Iowa State College, Kansas State College, Loyola University of Chicago, Marquette University, Mayo Foundation, Michigan College of Mining and Technology, Michigan State College, Northwestern University, University of Missouri, University of Nebraska, University of Pittsburgh, Notre Dame University, Ohio State University, Oklahoma A & M, Purdue University, St. Louis University, University of Chicago, University of Cincinnati, University of Illinois, University of Iowa, University of Kansas, University of Michigan, University of Minnesota, University of Wisconsin, Washington University of St. Louis and Western Reserve University.

At Brookhaven National Laboratory on Long Island, nine universities of the Northeast are participating in atomic research programs as members of Associated Universities, Inc., operator of the Laboratory. They are Columbia University, Cornell University, Harvard University, Johns Hopkins University, Massachusetts Institute of Technology, Princeton University, University of Pennsylvania, University of Rochester and Yale University.

Other important atomic research facilities are Ames Laboratory (Iowa State College), the University of California

Radiation Laboratory, the Knolls Atomic Power Laboratory, Schenectady, N. Y.; and the University of Rochester, Rochester, N. Y.

Vigorous comprehensive programs in all fields of atomic energy are being pursued in many different places under the auspices of the Atomic Energy Commission. Production of vital materials has been increased, research programs have been expanded and great strides have been made in the development of nuclear reactors for the ultimate production of useful power. In tests of atomic weapons in the Pacific—at Bikini in July 1946 and at Eniwetok in April 1948—new and important facts were learned concerning the national defense and security.

Resignations of four members of the original five-man Commission occurred in 1949 and 1950. First to leave was Mr. Waymack. His resignation was followed by that of Dr. Bacher, and on November 23, 1949, Mr. Lilienthal ended 20 years of public service by resigning as Commission chairman. Mr. Strauss left the Commission April 15, 1950. In a letter to President Truman, Mr. Lilienthal said:

"Although this resignation does bring my work as a public servant to a close, it does not mean that I do not intend to continue to be active in public affairs, for this is the first responsibility of all citizens in a democracy. Indeed, one of my chief reasons for wishing to return to private life is that I may be able to engage in public discussion and public affairs with a greater latitude than is either feasible or suitable for one who carries public responsibility."

"You have indeed," the President replied in accepting the resignation, "through almost 20 consecutive years of public service in tough pioneering jobs—always under tremendous pressure—earned the right to retire to private life."

Named as successors in the summer of 1949 to Mr. Waymack and Dr. Bacher were Gordon Dean, Professor of Law at
the University of California, who served as Assistant to Su-
preme Court Justice Jackson, United States Chief of Counsel
for prosecution of the major Nazi war criminals; and Dr.
Henry DeWolf Smyth, Chairman of the Department of Physics at Princeton University, and author of the famous "Smyth
Report," the historic technical document on the development
of atomic energy released August 12, 1945, under War Department auspices.

(Of the document, which became a controversial issue as to
the extent it furnished pertinent information for the development of atomic weapons, Dr. Smyth told the Joint Congressional Committee on Atomic Energy May 12, 1949: "I would
hate to be the director of a project that was handed that report and asked to make atomic bombs on the basis of it, because I don't believe I would get very far.")

In 1950, President Truman named Mr. Dean as Commission chairman and Mr. Pike, Dr. Smyth, Thomas E. Murray,
New York industrial engineer, and T. Keith Glennan, president of Case Institute of Technology, Cleveland, Ohio, as
other Commission members.

Atomic Energy's First Great Gift

Radioactive isotopes have been described as the most useful
research tool since the invention of the microscope in the 17th
Century. As atomic energy's first significant contribution to
the development of peacetime welfare, they have become ex-
ceedingly useful in biological, medical, agricultural and industrial research. They are speeding the fight against disease;
they are enabling scientists to gain a better understanding of

the working of the human body and they are helping man to make more efficient use of nature's materials, to grow more food and to produce better goods.

Radioactive isotopes assist science as sources of radiation for many important uses, including the treatment of disease, and as "tracers" of biological and physical processes formerly difficult or impossible to observe. Isotopes, already at work in hundreds of laboratories and hospitals in the United States and abroad, are adding much to man's knowledge about himself and the world around him.

What is an isotope and how is the isotope of an element made radioactive?

The word isotope is derived from two Greek words: "iso," meaning same, and "topos" meaning place. Isotopes are atoms; in fact the word isotope is merely used to identify different kinds of atoms of the same element. Most elements have families of atoms. The element carbon, for example, has four members in its family, carbon 11, carbon 12, carbon 13, and carbon 14. Some elements have more, some less. Each member of a particular element is called an isotope of that element and it can be distinguished by its individual and characteristic weight. In the case of carbon, the isotopes carbon 12 and carbon 13 occur in nature and are stable, whereas carbon 11 and carbon 14 are "man-made" and radioactive. Because carbon 11 and 14 are radioactive they will emit radiations by which scientists can also detect their presence. All isotopes of the same element, regardless of whether they are radioactive or stable, behave chemically in the same way. Because this is so and because some isotopes, the radioactive kind, can be identified by the radiation which they give off, they can be mixed with ordinary stable isotopes of the element and used to trace the atoms of this particular element through any chemical, biochemical or physical process. The identity and location of the radioactive isotope, invisible but potent, is determined by sensitive instruments.

Most elements, that is, the stable isotopes of the element, can be made radioactive by bombarding them with tiny particles called neutrons. Millions upon millions of these neutrons are produced every second in a nuclear reactor or "atomic furnace" by the splitting or fissioning of uranium-235 atoms. In the reactor at the Oak Ridge National Laboratory, the source of most of the radioactive isotopes being made available, about a million times a million neutrons are formed every second in a space the size of your little fingernail. For this reason, a nuclear reactor is the best device yet found for making radioactive isotopes.

The atomic rays in the "atomic furnace" are confined by a huge, protective, concrete shield which surrounds the "furnace," which is a huge "pile" of graphite in which uranium slugs of pre-determined size are arranged in a certain manner to bring about a chain reaction. If a sample of some stable element such as phosphorus, copper, zinc or iron is inserted in the "furnace" to be bombarded by the rays, they are made radioactive, or teeming with radiation.

This is what takes place. A sample of a stable element is placed in a small aluminum can. This is then inserted in a graphite carrier and shoved through a special opening into the "furnace." The whole operation may be likened to putting biscuits into an oven to cook—in this case an "atomic oven." The sample is "cooked"—actually irradiated by the intense breaking up of the uranium-235 atoms in the chain reaction—for some specified length of time, usually a week to over a month. It is now radioactivated; the result is a radioactive isotope from the original stable element placed in the "furnace."

The fact that radioisotopes of most of the elements can be

made in the uranium chain-reactor has made possible a wide scope of use in such fields as medical therapy, animal and plant physiology, bacteriology, physics, chemistry, industrial and agricultural research and metallurgy.

In the medical field, radioactive iodine is being used to treat persons with overactive thyroid glands, or with cancer of the thyroid, because the gland absorbs almost all the iodine taken into the body. Radioiodine is similarly being used to trace out new cancers that have spread from a thyroid root. Radiophosphorus is being used to curb the excessive production of red blood corpuscles in patients with the disease polychthemia vera and has gained relief for patients with leukemia, a disease marked by the cancerous overproduction of white blood corpuscles.

Medical workers are using radioactive iodine in tagged radioactive dye (diiodofluorescein) to locate brain tumors before surgery. The dye is taken up more selectively by brain tumor tissue than by normal brain tissue. The radioactive iodine's radiations penetrate the skull of the patient's head and show the surgeon where the tumor mass is located. Some investigators also are using radioactive phosphorus or P 32 as a supplementary tool in brain tumor surgery. The radiation from P 32 cannot penetrate through the skull and therefore cannot be used in the same way. If the patient is given P 32 before surgery, however, it also is selectively absorbed by brain tumor tissue. After the surgical incision has been made the surgeon can insert a small Geiger-counter tube (about one-eighth inch in diameter) through the incision into the mass of the brain and determine by the concentration of radioactivity what part of the tissue is tumor.

Metallic cobalt, when irradiated, emits radiations very similar to those of radium Radiocobalt is being used as a substitute for radium, one of the established tools for destroying

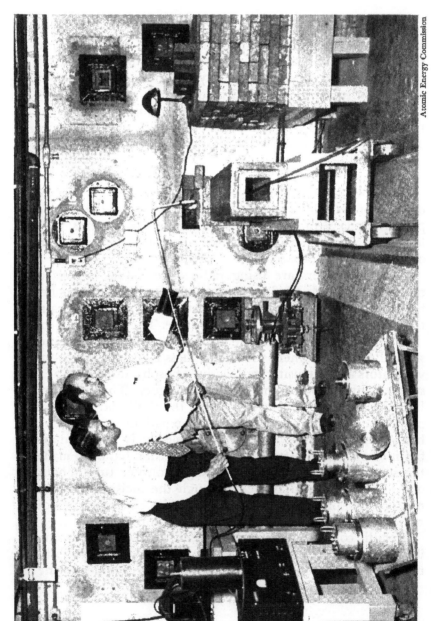

Magical radioactive elements from Oak Ridge's atomic "furnace" . . .

cancer cells. Radiocobalt, much cheaper in price than radium, can be made up in various pliable shapes having more potent radiation than radium and can be stored with proper protection as a fine wire on a spool and clipped off in quantities as desired. Irradiated cobalt plaques equivalent to million-volt X-ray machines have been developed. A powerful new tool is now available in the fight against cancer.

These uses are indicative of the wide value of the new research tool—radioactive materials. The first shipment of radioactive materials for research and medical purposes left the Oak Ridge National Laboratory August 2, 1946, almost a year from the time the first atomic bomb used in warfare fell on Hiroshima. Since that time, thousands upon thousands of shipments of radioactive isotopes of various elements have gone to research institutions and hospitals in this country and abroad.

An isotope helped make history; uranium-235, used in the first atomic bomb, is one of the isotopes of uranium. From the constructive uses of the new "man-made" radioactive isotopes will unfold new knowledge which will give comfort to all mankind.

American Museum of Atomic Energy

The American Museum of Atomic Energy, the world's first permanent museum devoted exclusively to telling the story of the atom, is situated in Oak Ridge. It is operated by the Oak Ridge Institute of Nuclear Studies, Inc.

Through scale models, pictures, maps and diagrams, the Museum shows, among other things, the massive plants for the separation of uranium-235, the model of a uranium chain-reacting pile (atomic "furnace"), the production of radioactive isotopes, the possibilities for production of power from nuclear energy, a 250,000-volt generator used in atomic energy work, a model of a pitchblende mine and the first small-scale model for the production of U-235 by the gaseous diffusion method. These and many other processes and principles of atomic energy are explained in exhibits occupying 18,000 square feet of floor space.

The Museum is operated for the benefit of the public.

CHAPTER X

Oak Ridge and the Future

CHANGE is the only tradition that Oak Ridge has known in its brief but dramatic existence.

Oak Ridge is taking deep root in Tennessee soil—green grass, rose bushes and well-kept lawns bespeak of community pride and permanency—but the changes continue. The transition is from a town wholly-owned and operated by the United States Government to one administered by its own citizens under established procedures of local self-rule—a self-governing and self-financing community.

An objective has been set for the city. It is for a community which will provide good living for people depended upon to do good work; it is to give Oak Ridge as great a degree of normality as can be achieved within the bounds of national defense and security and the efficient operation of the program it is built to serve—leadership in all fields of atomic energy.

In 1948, the Atomic Energy Commission set its sights on three major steps to make Oak Ridge a self-governing, self-supporting municipality occupying a normal place among other cities and towns of Tennessee.

They were (1) the elimination of barriers guarding Oak Ridge and the opening of the community proper to the general public with a simultaneous elaboration of security measures for the atomic energy production and research facilities; (2) private ownership of real property within Oak Ridge, either on the basis of sale or long-term lease of land, thereby stimulating development of needed facilities by private enterprise and stimulating home ownership; and (3) incorporation of the city under Tennessee charter.

The first of these objectives has been accomplished. The second step has been developed. Until the third problem is reconciled, Oak Ridge's city functions—police, fire and health departments, schools, street work and repairs and other activities—will continue to be administered-by city officials appointed and employed by the Atomic Energy Commission. The city has no tax structure, nor many other municipal features of the normal American community. Incorporation of the city is the prime requisite in this direction.

From the Spring of 1943 to March 1949, the entire Oak Ridge Area was closed. All entrances to the Area were restricted and passes to enter any portion of the Area, including the town itself, were necessary. For all intents and purposes, Oak Ridge was a forbidding, isolated place for the general American public.

But at 8:47 A.M. on Saturday, March 19, 1949, the community of Oak Ridge entered a new era. At that moment, in a ceremony at Elza Entrance—near the spot where five men stood on September 19, 1942, and made definite selection of a site for the building of secret atomic energy plants—a signal was given for the opening of the city. The signal was a small mushroom of smoke as an electrical impulse generated in the uranium chain-reactor at Oak Ridge National Laboratory 13 miles to the Southwest ignited a tape which had been treated with potassium chlorate and magnesium.

From 1943 to 1949, a total of 32,243,000 passes had been

[128]

A miniature "atomic mushroom" . . .

M. E. Wimberley

. . . and a huge parade signaled the opening of the town of Oak Ridge to the public March 19, 1949

Lillian Stokes Chastagner

written for admittance of persons into the city; now passes were no longer required. Of the Oak Ridge Area's 58,762 acres, 24,000, including the community, were opened for access to the public through Elza, Edgemoor, Solway and Oliver Springs entrances. New security fences were built to protect the new controlled area of 34,762 acres, not open to the general public, in which stand atomic energy production and research units.

Seventy-five thousand persons crowded into the city on March 19 for the festivities. They heard Vice-President Alben Barkley take note of the unique "opening" of an American town to warn that "the United States does not propose to be fenced in by any nation in the world"; they heard Senator Brien McMahon of Connecticut describe the event as "contrary to the current of the times; elsewhere in the world, areas of freedom are contracting, while here, in the least likely of places, the boundaries of freedom are expanding"; and they heard David E. Lilienthal, then Chairman of the Atomic Energy Commission, declare that "it is not natural and normal for Americans to live behind barriers; guards and fences and secret stamps on documents in themselves do not promote real security."

They heard not only these men, but others, including Senator Estes Kefauver, Governor Gordon Browning and Congressmen Albert Gore and John Jennings, Jr., all of Tennessee, and mingled with movie stars, among them Lee Bowman, Marie McDonald, Rod Cameron and Adolphe Menjou, who expressed the hope that "never again in the history of mankind will it be necessary for Oak Ridge to unleash the genii of atomic energy for the destruction of mankind."

For Oak Ridge, it was the biggest day since August 6, 1945. Being different—living behind wire fences and guards—had become tiresome for most of the residents. The fears of a few dissenters that the opening of the city would increase crime, encourage prowlers, and endanger children through non-observance of safety rules by an influx of visitors, have not been realized.

Many factors are giving impetus to the movement of Oak Ridge toward normality. Serving the community is a five-day-a-week afternoon newspaper, *The Oak Ridger,* and a radio station, WATO, both independently owned and operated. A Chamber of Commerce is active; the city has a Golf and Country Club.

Hundreds of new homes and apartments to replace emergency shelter built during wartime have been constructed, as have new schools. The site for a new central business district, on which private businesses have been encouraged to erect their own buildings, has been developed. A Master Plan for the orderly transition of the city from its camptown atmosphere to one having permanent aspects has been used as a guide for the city's future physical growth.

Land already has been sold to churches for construction of their own buildings. Two funeral homes are in operation but so far no cemeteries for use by Oak Ridge residents have been set aside; the 65 cemeteries in the Oak Ridge Area are those taken over by the Government in 1942. Oak Ridge has no railroad station or airport, but these have been discussed. Management of residential and business property remains in Government hands but it is anticipated that private interests will eventually handle all the properties.

When World War II ended, opinions differed as to what had been hatched in Oak Ridge besides the material for the atomic bomb. Some felt it was a "hellish" place, others regarded it as "heavenly."

Those sharing the latter view subscribed to the prediction made around the turn of the Twentieth Century by John Hendrix, the prophet of the East Tennessee hills, that the great city he saw in his visions and which he said would be built on the ridges and the valleys now constituting the Oak Ridge Area would be known as "Paradise."

But Oak Ridge is no "paradise." The people who make it up are the same as those who make up the thousands of other American cities and towns.

The future of Oak Ridge depends on many minds and hearts. On September 23, 1949, President Truman announced to the American public that "we have evidence that within recent weeks an atomic explosion occurred in Russia."

With agreement on international control of atomic energy still undetermined on that date, the immediate reaction of the United States Government was the authorization of construction of giant, new facilities for the production of fissionable materials at Oak Ridge involving an expenditure of millions of dollars. Research facilities also have been strengthened.

Then, on January 31, 1950, President Truman, in a world-shaking announcement, revealed that "I have directed the Atomic Energy Commission to continue its work on all forms of atomic weapons, including the so-called hydrogen, or super-bomb . . . until a satisfactory plan for international control of atomic energy is achieved." Congress appropriated millions for the project and a site other than Oak Ridge was selected for the building of plants for production of certain materials.

So, of Oak Ridge, the city, and of Oak Ridge, the atomic center, at least one accurate prediction can be made:

Oak Ridge will never be an entirely normal place. For not

only are the 30,000 persons who make up its population concerned with it but 150,000,000 other Americans as well. Oak Ridge can never live unto itself alone.

Safety in the Atomic Age

From its inception, Oak Ridge's record in all fields of safety has been remarkable. Despite the great hazards of radiation during the building and operation of the atomic energy facilities, there have been no known overexposures to radiation in Oak Ridge. The Manhattan District program of protection against radiation hazards was developed by Col. Stafford L. Warren, chief of the District's Medical Division, and Dr. Robert S. Stone, of the University of California, who served as Associate Project Director for Health. The program has been carried forward and improved on by the Atomic Energy Commission, so successfully that in 1949 a representative of the insurance industry said that "from the standpoint of health and safety, the records of the atomic plants and laboratories are excellent. It is as safe to work in any atomic energy plant as it is to work in any heavy or chemical industry."

On December 5, 1945, the Manhattan District and its contractors were presented the National Safety Council's Award for Distinguished Service to Safety for "achieving and maintaining low accident rates at the Manhattan District facilities throughout the country under the urgent demands for speed in the unique processes attaining the development of the atomic bomb and thereby making a signal contribution to early victory."

Oak Ridge, the city, won the National Safety Council's Award of Merit Plaque in 1945 for its record of traffic safety. In 1946, 1947 and 1948, the city received the National Safety Council's Citation for Achievement in traffic safety. An

The present-day Oak Ridge sparkles with new schools . . . .

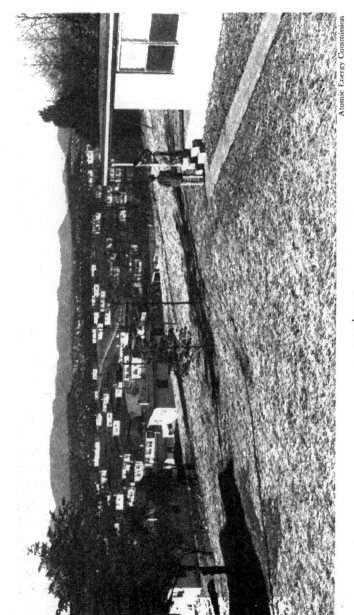

. . . new homes . . .

Atomic Energy Commission

. . . and new apartments . . .

. . . of modern design . . .

. . . as it travels . . .

. . . toward a new era

Award of Recognition from the Tennessee Safety Council also went to the city in 1948.

At 3:15 A.M. on November 5, 1949, one of the most impressive records in the history of American safety campaigns —for cities of over 15,000 population—1,423 days without a traffic fatality in Oak Ridge—came to an end with the death of an automobile driver as his car left the fog-shrouded Oak Ridge Turnpike and overturned.

In 1945, 1946, 1947, 1948 and 1949, Honor Roll Certifications from the National Safety Council went to the eleven schools in Oak Ridge for their safety accomplishments.

In 1949, American Industrial Transport, Inc., operators of the Oak Ridge bus system, was awarded first place in the nation by the National Safety Council for its safety record in the city suburban division contest for the period July 1, 1948, to June 30, 1949. American Industrial Transport also placed first in safety in 1947 in Group One, City Buses Division, National Safety Council contest and was second in the same division in 1948.

The city also has received several awards in fire prevention campaigns, placing first in 1947 and 1948 in the United States and Canada in contests sponsored by the National Fire Prevention Association. In 1948, the city also received a Certificate of Merit in the Fire Waste Contest sponsored by the United States Chamber of Commerce.

The K-25 and Y-12 plants of Carbide and Carbon Chemicals Division tied for first place in 1948 in the National Industrial Competition of the National Fire Prevention Association and Roane-Anderson Company, management contractor for the community, placed fourth.

Appendices

Organizations Receiving Army-Navy "E" Awards.

———

Personnel in Manhattan District Receiving Significant War Department Awards.

———

An Accounting of the World's First Atomic Bomb Blast.

———

Scientific Developments Leading to the Formation of the Manhattan District and the Building of Oak Ridge.

———

The Provision of Manpower for the Manhattan Project.

———

Tribute to Manhattan District Personnel.

———

An Accounting of the Successful Operation of the First Self-Sustaining Nuclear Chain Reactor in World History.

———

Organizations and Churches in Oak Ridge.

[134]

"E" Awards

In a ceremony at Oak Ridge September 29, 1945, at which Secretary of War Robert Patterson was the principal speaker, the following contractors who participated in the building and operations of Oak Ridge facilities were presented the Army-Navy "E" Award for "excellence in war production":

Carbide and Carbon Chemicals Division, Union Carbide
  and Carbon Corporation
Tennessee Eastman Corporation
Stone & Webster Engineering Corporation
Monsanto Chemical Company
H. K. Ferguson Company
The Fercleve Corporation
Ford, Bacon and Davis, Inc.
J. A. Jones Construction Company
Kellex Corporation

Other organizations which also received "E" Awards as participants in the Manhattan District program were:

Hooker Electro-Chemical Company, Niagara Falls, N. Y.
Electro Metallurgical Company, Niagara Falls, N. Y.
Linde Air Products Company, Buffalo, N. Y.
Harshaw Chemical Company, Cleveland, Ohio
Chrysler Corporation, Detroit, Michigan
Houdaille-Hershey Company, Decatur, Ill.
Mallinckrodt Chemical Works, St. Louis, Mo.
Iowa State College, Ames, Iowa
Scientific Laboratory, University of California,
  Los Alamos, New Mexico
McKee Construction Company, Los Alamos, New Mexico
E. I. du Pont de Nemours and Company, Richland,
  Washington

Certificates of Merit signed by the Secretary of War also were presented the University of California, the University of Chicago and Columbia University.

Awards to Army Personnel

Special War Department recognition went to the following Army personnel in the Fall of 1945 for their work on the A-Bomb Project:

Distinguished Service Medal

General Leslie R. Groves (Washington)
General Thomas F. Farrell (Washington)
Col. Kenneth D. Nichols (Oak Ridge)
Col. Stafford L. Warren (Oak Ridge)
Col. Franklin T. Matthias (Hanford)

Legion of Merit

(Oak Ridge personnel)

Col. Walter J. Williams
Lt. Col. Richard W. Cook
Lt. Col. Charles Vanden Bulck
Col. Earl H. Marsden
Col. W. B. Parsons
Lt. Col. John S. Hodgson
Lt. Col. John R. Ruhoff
Lt. Col. Arthur V. Peterson
Lt. Col. Hymer L. Friedell
Lt. Col. Charles E. Rea
Lt. Col. Mark C. Fox
Lt. Col. Alfonso Tammaro
Lt. Col. W. P. Cornelius
Lt. Col. Curtis A. Nelson
Major Harry S. Traynor

Capt. Arlene G. Scheidenhelm
Warrant Officer Murray S. Levine
Lt. George O. Robinson, Jr.

Legion of Merit

(Other Manhattan District personnel)

Col. Donald E. Antes
Col. Clarence D. Barker
Major H. K. Calvert
Col. William A. Consodine
Col. John A. Derry
Lt. Col. Peer de Silva
Major Harold A. Fidler
Lt. Col. Allan C. Johnson
Major Wilbur E. Kelley
Col. John Lansdale, Jr.
Brig. Gen. James C. Marshall
Capt. James F. Nolan
Lt. Walter A. Parish
Major Claude C. Pierce, Jr.
Col. Lyle E. Seeman
Major Francis J. Smith
Lt. Col. Stanley L. Stewart
Lt. Col. James C. Stowers
Col. Gerald R. Tyler
Major John E. Vance
Lt. Joseph Volpe, Jr.
Major Robert J. Wier

*An Accounting of the World's First Atomic Bomb Blast,
Alamogordo, New Mexico, July 16, 1945, Official
War Department Release—August 6, 1945*

Mankind's successful transition to a new age, the Atomic
Age, was ushered in July 16, 1945, before the eyes of a tense

[137]

group of renowned scientists and military men gathered in the desertlands of New Mexico to witness the first end results of their $2,000,000,000 effort. Here in a remote section of the Alamogordo Air Base 120 miles southeast of Albuquerque the first man-made atomic explosion, the outstanding achievement of nuclear science, was achieved at 5:30 A.M. of that day. Darkening heavens, pouring forth rain and lightning immediately up to the zero hour, heightened the, drama.

Mounted on a steel tower, a revolutionary weapon destined to change war as we know it, or which may even be the instrumentality to end all wars, was set off with an impact which signalized man's entrance into a new physical world. Success was greater than the most ambitious estimates. A small amount of matter, the product of a chain of huge specially constructed industrial plants, was made to release the energy of the universe locked up within the atom from the beginning of time. A fabulous achievement had been reached. Speculative theory, barely established in pre-war laboratories, had been projected into practicality.

This phase of the Atomic Bomb Project, headed by Major General Leslie R. Groves, was under the direction of Dr. J. R. Oppenheimer, theoretical physicist of the University of California. He is to be credited with achieving the implementation of atomic energy for military purposes.

Tension before the actual detonation was at a tremendous pitch. Failure was an ever-present possibility. Too great a success, envisioned by some of those present, might have meant an uncontrollable, unusable weapon.

Final assembly of the atomic bomb began on the night of July 12 in an old ranch house. As various component assemblies arrived from distant points, tension among the scientists rose to an increasing pitch. Coolest of all was the man charged with the actual assembly of the vital core, Dr. R. F. Bacher, in normal times a Professor at Cornell University.

The entire cost of the project, representing the erection of whole cities and radically new plants spread over many miles of countryside, plus unprecedented experimentation, was represented in the pilot bomb and its parts. Here was the focal point of the venture. No other country in the world had been capable of such an outlay in brains and technical effort.

The full significance of these closing moments before the final factual test was *not* lost on these men of science. They fully knew their position as pioneers into another Age. They also knew that one false move would blast them and their entire effort into eternity. Before the assembly started a receipt for the vital matter was signed by Brigadier General Thomas F. Farrell, General Groves' deputy. This signalized the formal transfer of the irreplaceable material from the scientists to the Army.

During final preliminary assembly, a bad few minutes developed when the assembly of an important section of the bomb was delayed. The entire unit was machine-tooled to the finest measurement. The insertion was partially completed when it apparently wedged tightly and would go no farther. Dr. Bacher, however, was undismayed and reassured the group that time would solve the problem. In three minutes' time, Dr. Bacher's statement was verified and basic assembly was completed without further incident.

Specialty teams, comprised of the top men on specific phases of science, all of which were bound up in the whole, took over their specialized parts of the assembly. In each group was centralized months and even years of channelized endeavor.

On Saturday, July 14, the unit which was to determine the

success or failure of the entire project was elevated to the top of the steel tower. All that day and the next, the job of preparation went on. In addition to the apparatus necessary to cause the detonation, complete instrumentation to determine the "pulse beat" and all reactions of the bomb was rigged on the tower.

The ominous weather which had dogged the assembly of the bomb had a very sobering effect on the assembled experts whose work was accomplished amid lightning flashes and peals of thunder. The weather, unusual and upsetting, blocked out aerial observation of the test. It even held_up the actual explosion scheduled at 4 A.M. for an hour and a half. For many months the approximate date and time had been set and had been one of the high level secrets of the best kept secret of the entire war.

Nearest observation point was set up 10,000 yards south of the tower where in a timber and earth shelter the controls for the tests were located. At a point 17,000 yards from the tower at a point which would give the best observation the key figures in the atomic bomb project took their posts. These included General Groves, Dr. Vannevar Bush, head of the Office of Scientific Research and Development, and Dr. James B. Conant, president of Harvard University.

Actual detonation was in charge of Dr. K. T. Bainbridge of Massachusetts Institute of Technology. He and Lieutenant Bush, in charge of the Military Police Detachment, were the last men to inspect the tower with its cosmic bomb.

At three o'clock in the morning the party moved forward to the control station. General Groves and Dr. Oppenheimer consulted with the weathermen. The decision was made to go ahead with the test despite the lack of assurance of favorable weather. The time was set for 5:30 A.M.

General Groves rejoined Dr. Conant and Dr. Bush and

just before the test time, they joined the many scientists gathered at the Base Camp. Here all present were ordered to lie on the ground, face downward, heads away from the blast direction.

Tension reached a tremendous pitch in the control room as the deadline approached. The several observation points in the area were tied in to the control room by radio and with 20 minutes to go, Dr. S. K. Allison of Chicago University took over the radio net and made periodic time announcements.

The time signals, "minus 20 minutes, minus fifteen minutes," and on and on increased the tension to the breaking point as the group in the control room which included Dr. Oppenheimer and General Farrell held their breaths, all praying with the intensity of the moment which will live forever with each man who was there. At "minus 45 seconds," robot mechanism took over and from that point on the whole great complicated mass of intricate mechanism was in operation without human control. Stationed at a reserve switch, however, was a soldier scientist ready to attempt to stop the explosion should the order be issued. The order never came.

At the appointed time, there was a blinding flash lighting up the whole area brighter than the brightest daylight. A mountain range three miles from the observation point stood out in bold relief. Then came a tremendous sustained roar and a heavy pressure wave which knocked down two men outside the control center. Immediately thereafter, a huge multi-colored surging cloud boiled to an altitude of over 40,000 feet. Clouds in its path disappeared. Soon the shifting substratosphere winds dispersed the now gray mass.

The test was over, the project a success.

The steel tower had been entirely vaporized. Where the

tower had stood, there was a huge sloping crater. Dazed but relieved at the success of their tests, the scientists promptly marshalled their forces to estimate the strength of America's new weapon. To examine the nature of the crater, specially equipped tanks were wheeled into the area, one of which carried Dr. Enrico Fermi, noted nuclear scientist.

Had it not been for the desolated area where the test was held and for the cooperation of the press in the area, it is certain that the test itself would have attracted far reaching attention. As it was, many people in that area are still discussing the effect of the smash. A significant aspect, recorded by the press, was the experience of a blind girl near Albuquerque.many miles from the scene, who, when the flash of the test lighted the sky before the explosion could be heard, exclaimed, "What was that?"

Interviews of General Groves and General Farrell give the following on-the-scene versions of the test. General Groves said: "My impressions of the night's high points follow: After about an hour's sleep I got up at 0100 and from that time on until about five I was with Dr. Oppenheimer constantly. Naturally he was tense, although his mind was working at its usual extraordinary efficiency. I attempted to shield him from the evident concern shown by many of his assistants who were disturbed by the uncertain weather conditions. By 0330 we decided that we could probably fire at 0530. By 0400 the rain had stopped but the sky was heavily overcast. Our decision became firmer as time went on.

"During most of these hours the two of us journeyed from the control house out into the darkness to look at the stars and to assure each other that the one or two visible stars were becoming brighter. At 0510 I left Dr. Oppenheimer and returned to the main observation point which was 17,000 yards from the point of explosion. In accordance with our orders

I found all personnel not otherwise occupied massed on a bit of high ground.

"Two minutes before the scheduled firing time, all persons lay face down with their feet pointing towards the explosion. As the remaining time was called from the loud speaker from the 10,000 yard control station there was complete awesome silence. Dr. Conant said he had never imagined seconds could be so long. Most of the individuals in accordance with orders shielded their eyes in one way or another.

"First came the burst of light of a brilliance beyond any comparison. We all rolled over and looked through dark glasses at the ball of fire. About forty seconds later came the shock wave followed by the sound, neither of which seemed startling after our complete astonishment at the extraordinary lighting intensity.

"A massive cloud was formed which surged and billowed upward with tremendous power, reaching the substratosphere in about five minutes.

"Two supplementary explosions of minor effect other than the lighting occurred in the cloud shortly after the main explosion.

"The cloud traveled to a great height first in the form of a ball, then mushroomed, then changed into a long trailing chimney-shaped column and finally was sent in several directions by the variable winds at the different elevations.

"Dr. Conant reached over and we shook hands in mutual congratulations. Dr. Bush, who was on the other side of me, did likewise. The feeling of the entire assembly, even the uninitiated, was of profound awe. Drs. Conant and Bush and myself were struck by an even stronger feeling that the faith of those who had been responsible for the initiation and the carrying on of this Herculean project has been justified."

General Farrell's impressions are: "The scene inside the

shelter was dramatic beyond words. In and around the shelter were some twenty odd people concerned with last minute arrangements. Included were Dr. Oppenheimer, the Director who had borne the great scientific burden of developing the weapon from the raw materials made in Tennessee and Washington, and a dozen of his key assistants, Dr. Kistiakowsky, Dr. Bainbridge, who supervised all the detailed arrangements for the test; the weather expert, and several others. Besides these, there were a handful of soldiers, two or three Army officers and one Naval officer. The shelter was filled with a great variety of instruments and radios.

"For some hectic two hours preceding the blast, General Groves stayed with the Director. Twenty minutes before the zero hour, General Groves left for his station at the base camp, first because it provided a better observation point and second, because of our rule that he and I must not be together in situations where there is an element of danger which existed at both points.

"Just after General Groves left, announcements began to be broadcast of the interval remaining before the blast to the other groups participating in and observing the test. As the time interval grew smaller and changed from minutes to seconds, the tension increased by leaps and bounds. Everyone in that room knew the awful potentialities of the thing that they thought was about to happen. The scientists felt that their figuring must be right and that the bomb had to go off but there was in everyone's mind a strong measure of doubt. The feeling of many could be expressed by 'Lord, I believe; help Thóu mine unbelief.'

"We were reaching into the unknown and we did not know what might come of it. It can safely be said that most of those present were praying and praying harder than they

[144]

had ever prayed before. If the shot were successful, it was a justification of the several years of intensive effort of tens of thousands of people—statesmen, scientists, engineers, manufacturers, soldiers, and many others in every walk of life.

"In that brief instant in the remote New Mexico desert, the tremendous effort of the brains and brawn of all these people came suddenly and startlingly to the fullest fruition. Dr. Oppenheimer, on whom had rested a very heavy burden, grew tenser as the last seconds ticked off. He scarcely breathed. He held on to a post to steady himself. For the last few seconds, he stared directly ahead and then when the announcer shouted "Now!" and there came this tremendous burst of light followed shortly thereafter by the deep growling roar of the explosion, his face relaxed into an expression of tremendous relief. Several of the observers standing back of the shelter to watch the lighting effects were knocked flat by the blast.

"The tension in the room let up and all started congratulating each other. Everyone sensed "This is it!" No matter what might happen now all knew that the impossible scientific job had been done. Atomic fission would no longer be hidden in the cloisters of the theoretical physicists' dreams. It was almost full grown at birth. It was a great new force to be used for good or for evil. There was a feeling in that shelter that those concerned with its nativity should dedicate their lives to the mission that it would always be used for good and never for evil.

"Dr. Kistiakowsky threw his arms around Dr. Oppenheimer and embraced him with shouts of glee. Others were equally enthusiastic. All the pent-up emotions were released in those few minutes and all seemed to sense immediately that the explosion had far exceeded the most optimistic ex-

[145]

pectations and wildest hopes of the scientists. All seemed to
feel that they had been present at the birth of a new age—
The Age of Atomic Energy—and felt their profound responsibility to help in guiding into right channels the tremendous forces which had been unlocked for the first time in
history.

"As to the present war, there was a feeling that no matter
what else might happen, we now had the means to insure its
speedy conclusion and save thousands of American lives. As
to the future, there had been brought into being something
big and something new that would prove to be immeasurably
more important than the discovery of electricity or any of the
other great discoveries which have so affected our existence.

"The effects could well be called unprecedented, magnifi-
cent, beautiful, stupendous and terrifying. No man-made
phenomenon of such tremendous power had ever occurred
before. The lighting effects beggared description. The whole
country was lighted by a searing light with the intensity
many times that of the midday sun. It was golden, purple,
violet, gray and blue. It lighted every peak, crevasse and
ridge of the nearby mountain range with a clarity and beauty
that cannot be described but must be seen to be imagined. It
was that beauty the great poets dream about but describe
most poorly and inadequately. Thirty seconds after, the explosion came first, the air blast pressing hard against the
people and things, to be followed almost immediately by the
strong, sustained, awesome roar which warned of doomsday
and made us feel that we puny things were blasphemous to
dare tamper with the forces heretofore reserved to The Almighty. Words are inadequate tools for the job of acquaint-
ing those not present with the physical, mental and psycho-
logical effects. It had to be witnessed to be realized."

[146]

Scientific Developments Leading to the Formation of the Manhattan District and the Building of Oak Ridge

The phenomenon of uranium fission, the fact that absorption of a neutron by a uranium nucleus (U-235) sometimes causes that nucleus to split into approximately equal parts with the release of enormous quantities of energy, was confirmed early in January 1939 in Denmark by Lise Meitner, a woman, and O. R. Frisch, both refugees from Hitler's Germany and colleagues of Dr. Niels Bohr, eminent physicist of Copenhagen. They confirmed experiments originally made by Otto Hahn and Fritz Strassman at the Kaiser Wilhelm Institute in Berlin.

Dr. Bohr, who at a later date in World War II was spirited from under the Nazis' noses in Denmark by British Intelligence and taken to England to assist in certain phases of atomic energy development, communicated the fact of uranium fission to a former student, J. A. Wheeler and others at Princeton University, while on a visit to the United States in the latter part of January 1939.

The news reached other physicists, including Dr. Enrico Fermi, Italian-born physicist, at Columbia University. Following conversations between Dr. Fermi, Dr. J. R. Dunning and Dr. G. B. Pegram, also of Columbia, arrangements were made for additional experiments.

Experiments at Columbia not only confirmed uranium fission and the implication of a chain reaction but simultaneous experimental confirmation came from other research groups at Carnegie Institution of Washington, Johns Hopkins University and the University of California.

The possible military importance of uranium fission was called to the attention of the Navy in a conference, arranged through Dr. G. B. Pegram in March 1939, by Dr. Fermi, who suggested the possibility of achieving a controllable uranium chain reaction for explosive purposes. The Navy asked to be kept informed on developments.

Dr. Leo Szilard and Dr. Eugene Wigner, also foreign-born physicists, made the next move to stimulate Government interest. In July 1939 they conferred with Dr. Albert Einstein. A short time later Drs. Einstein, Szilard and Wigner discussed the problem with Alexander Sachs, a New York economist and friend of Roosevelt. That Fall, Mr. Sachs, supported by a letter from Dr. Einstein, explained to President Roosevelt the urgency of the matter. The President then appointed an "Advisory Committee on Uranium" consisting of L. J. Briggs as chairman, Colonel K. F. Adamson of the Army Ordnance Department and Commander G. C. Hoover of the Navy Bureau of Ordnance.

First funds for the purchase of necessary materials—graphite and uranium oxide—for certain measurements was a sum of $6,000 from the Army and Navy. When the Committee next met on April 28, 1940, new factors pointing to successful use of atomic energy for bombs had been developed and arrangements were already underway with the Union Miniere of the Belgian Congo looking toward obtaining a large supply of uranium ore.

With the organization of the National Defense Research Committee in June 1940, President Roosevelt directed that the Uranium Committee be reconstituted as a subcommittee of the NDRC, reporting to Dr. Vannevar Bush, NDRC chairman. First contracts for research from the NDRC went to Columbia University, where in time the first U-235 was obtained through the gaseous diffusion process on a laboratory scale. (This laboratory model now rests in the American Museum of Atomic Energy at Oak Ridge.)

Other research contracts went to Princeton University, Standard Oil Development Company, Cornell University, Carnegie Institution of Washington, Johns Hopkins University, Massachusetts Institute of Technology, University of Virginia, National Bureau of Standards, the University of Chicago, the University of California, and the University of Minnesota, where Dr. A. O. Nier first obtained U-235 through the mass spectrograph method, which stimulated Dr. E. O. Lawrence of the University of California to develop the electromagnetic separation of U-235, feasibility of which was confirmed by Lawrence on December 8, 1941, the day after Pearl Harbor.

By November 1941, $300,000 had been allotted for research. In December, with interchange of information with the British already underway, Dr. Bush and associates felt that the possibility of obtaining atomic bombs was great enough to justify an "all out" effort.

This led to the formation of a separate organization, the Office of Scientific Research and Development. At a meeting December 16, 1941—nine days after Pearl Harbor—of the Government's Top Policy Committee composed of Vice-President Henry A. Wallace, Secretary of War Henry L. Stimson and Dr. Bush (General George C. Marshall and Dr. J. B. Conant of Harvard, other members, were absent), decisions were reached which gave great impetus in 1942 to research on the gaseous diffusion, electromagnetic and thermal diffusion methods for obtaining U-235 and the uranium chain-reactor program for obtaining plutonium.

On June 13, 1942, a report recommended expansion of the atomic bomb program. President Roosevelt approved it on June 17. On June 18 Colonel J. C. Marshall, Corps of Engineers, was instructed by the Chief of Engineers to form a new Engineer District to carry on special work (atomic bombs) in behalf of the United States, Britain and Canada and their allies. Great supplies of uranium ore were to be furnished by Canada from its Great Bear Lake region.

Formation of the Manhattan District followed on August 13, 1942. On September 17, 1942, Secretary of War Stimson placed Brig. Gen. L. R. Groves of the Corps of Engineers in complete charge of all Army activities relating to the secret DSM (Development of Substitute Materials) Project.

The Provision of Manpower for the Manhattan Project, Official War Department Release—August 6, 1945

Once the magnitude of the atomic bomb project had been established, manpower immediately was recognized as one of the key ingredients which would spell the difference between success or failure. The Army was faced with its two largest construction jobs, the largest in modern times and possibly the largest in history. In addition to the usual obstacles, a stepped-up schedule had to be met, time being of the essence in a grim race against the unknown schedule of the Germans.

The project had an unusual obstacle to face. Security was paramount. At this time, national competition for manpower was acute. Industries and war projects were vieing with each other in this competition, citing the key part their people were playing in the war effort. No such inducement could be made to attract labor to the Atomic Bomb Project. Nothing whatsoever could be told in recruitment beyond the fact that the work would be in the top interests of the war endeavor.

At first, the general attitude was that the project's construction was just another job—or that "business as usual" was the order of the day. Trade unions, the War Manpower Commission, plus the Manhattan District's expediters, teamed to achieve what at times seemed impossible provision of adequate manpower. Heading this program was Colonel Clarence D. Barker, chief of the Labor Division of the Office of the Chief of Engineers.

By the time the Manhattan District began its large scale recruiting activities, the War Manpower Commission and its agencies were well established and labor recruiting was carried on primarily through their services. The U. S. Employment Service utilized the American Federation of Labor to recruit and move skilled tradesmen. The common laborers' union, however, did not have sufficient membership to supply demands and these were recruited through the U. S. E. S. from the general labor market.

Types of personnel necessary to man the project covered practically all occupational skills. These ranged from common laborers, carpenters and plumbers to glass blowers, chemists and physicists. The mass of personnel, however, fell into two general classes: construction laborers and mechanics and plant operators.

Recruitment of special skills such as chemists, physicists, laboratory technicians and others presented many problems. As a whole, they were as difficult to find as the larger numbers of the more common skills. The most difficult problems in this phase were handled personally by Dr. Samuel Arnold, Dean of Men at Brown University, himself an eminent scientist.

Much of the supervisory and technical personnel were recruited by the many contractors of the Manhattan District within their own organizations. Many of the top scientists were brought to the project through contracts placed with various universities.

The recruiting of operations people was particularly a difficult problem because of the necessity of training all new people for the work. It necessitated the stripping of the operating contractors of a great many of the key men of their organizations which in view of the increased activities brought on by the war programs other than that connected with the Manhattan District had made the situation more complex.

This was the overall personnel procurement program of the Manhattan Engineer District. But there were many problems which at times seemed to defy solution. Had it not been for the complete coordination of the whole problem, several situations could have progressed to disastrous proportions.

The construction, by reasons of its immensity and uniqueness and also because of a great many new practices developed which had never been used in the industry before necessitated the support of the top labor leaders. On several occasions it was necessary that Judge Robert Patterson, the Under Secretary of War call in the leaders, including the President of A. F. of L., Mr. William Green and the General Presidents of several Building Trades Unions to seek their cooperation and to give them a better understanding of the problems involved.

They, in a great many instances, broke down conditions of long standing in order that the completion *on* schedule be not interfered with. Judge Patterson also gave a great deal of his personal time to this phase when it was required.

By June 15, 1944, the shortage of electricians at the Hanford Engineer Works, Washington, and the Clinton Engineer Works, Tennessee, had become so acute that work schedules were seriously endangered. Twenty-five hundred electricians had to be recruited. A plan was worked out by the Under Secretary of War and Edward J. Brown, president of the International Brotherhood of Electrical Workers. Electricians would be borrowed from other employers for a period of 90 days. The National Electrical Contractors Association was called in and a carefully worded news release for security reasons was issued by the War Department stating the project's predicament. In two months' time, the bottleneck was completely and satisfactorily broken. The plan was continued throughout construction.

An acute shortage of machinists and toolmakers late in 1943 resulted in stringent measures. The New Mexico installation urgently needed 190 men in these skills. The War Manpower Commission issued instructions to its regional directors on October 21, 1943, authorizing them to certify certain workers as available to the Manhattan District even over the protests of their employers, many of whom were in other essential war programs. With this authority as a basis, special recruiting teams composed of an Army officer, a recruiter, and a security agent procured the workers needed in one month.

The Manhattan District experienced more unusual problems of turnover and absenteeism than other war industries and installations. This was directly due to the isolation of the projects, the extended length of the construction period, expansions in the construction program, security, and limited housing and crowded transportation facilities.

A rigorous campaign was set up to solve these problems. Exit interviews salvaged many. In hundreds of cases, competent employees were either persuaded to go back to work or to take other jobs on the same project. Employees made available by reduction in force were also picked up in this manner and directed to other jobs on the project or in some cases returned to essential industry. These interviews also determined why workers were leaving and set up a basis for corrective action.

Companion problem to turnover was absenteeism. Repeated absenteeism was the greatest single cause for terminations. War economy with its larger incomes resulting from higher wages and longer hours provided less compulsion for steady work than the lower incomes of peace time. Therefore, every effort was made, within the limits of the isolated areas where the projects were established, to better living and working conditions.

It was soon found that job dissatisfaction as a whole hinged on lack of facilities present in normal American communities. To the seasoned construction worker, conditions were average. To the men having their first fling at construction and to the men and women who took production jobs, life was markedly different. The Army attempted to make conditions more normal by providing recreation facilities as movie houses, baseball diamonds, tennis courts and recreation halls. These facilities greatly assisted in keeping workers on the job.

The Army also provided subsidized transportation, nursery schools for working mothers, tire and gasoline rationing boards and conveniently located shopping facilities.

The following unions were associated with the construction phases of the project:

Int'l Ass'n of Heat and Frost Insulators and Asbestos Workers
Int'l Brotherhood of Boiler Makers, Iron Ship Builders and Helpers
Bricklayers, Masons and Plasterers' Int'l Union
United Brotherhood of Carpenters and Joiners
Int'l Brotherhood of Electrical Workers
Int'l Union of Elevator Constructors
Int'l Union of Operating Engineers
Int'l Ass'n of Bridge, Structural and Ornamental Iron Workers
Int'l Hod Carriers, Bldg., and Common Laborers Union

Wood, Wire and Metal Lathers' Int'l Union
Brotherhood of Painters, Decorators and Paperhangers
Operative Plasterers and Cement Finishers Int'l Ass'n
United Slate, Tile and Composition Roofers, Damp & Water-
proof Workers Ass'n
United Ass'n of Journeymen Plumbers and Steam Fitters
Sheet Metal Workers' International Ass'n
Int'l Brotherhood of Teamsters, Chauffeurs, Warehousemen
and Helpers
Building Trades Dept. of AFL
Int'l Ass'n of Machinists
Int'l Brotherhood of Firemen and Oilers

Tribute to Manhattan District Personnel, Official War Department Release—August 9, 1945

OAK RIDGE, Tenn.—Col. Kenneth D. Nichols, District Engineer of the Manhattan Engineer District, has paid tribute to the hundreds of organizations and thousands of persons whose coordinated efforts have made possible the utilization of atomic power within the time span of this war for use in bombs against Japan.

Explaining that the progress of fundamental research in physics and chemistry prior to the war indicated that utilization of atomic power might have been feasible in 15 to 20 years, Colonel Nichols declared that the combined effort of the many different people and organizations connected with the project has compressed the time to three years, an accomplishment which will endure as a monument to the ingenuity and vision and determination of all those, from scientists to laborers, who have had a part in the work.

"These people and organizations—scientific, engineering, contracting, manufacturing, procuring and others—working in harmony among themselves and with Government agencies deserve unlimited credit for the successful accomplishment of an almost impossibly vast and complicated task," Colonel Nichols declared.

In addition, he pointed out that the District's staff of specially selected officers, WACs, enlisted men, and civilians deserve a large measure of credit for the success of the Army's part in the project. Paying tribute to the work they have done, Colonel Nichols declared that "each assistant has spent long hours of work each day and collectively have made it possible for the Manhattan District to control the large volume of research, construction, and production necessary to complete the project."

Colonel Nichols said it is impossible to list the several hundred names of military and civilian personnel assigned to the Manhattan District, but added that he wished to mention a few of those who "have made exemplary contributions to the success of the project."

Among these, he said, are:

Col. E. H. Marsden, of New Haven, Conn., Executive Officer for the District; Col. F. T. Matthias, of Curtis, Wisconsin, Area Engineer for the Hanford Engineer Works at Richland, Wash.; Col. G. R. Tyler, of Philadelphia, Pa., Area Engineer at Santa Fe; and Col. S. L. Warren, of Rochester, N. Y., Chief of the Medical Section, who was formerly Professor, School of Medicine and Dentistry, University of Rochester.

In addition, Colonel Nichols said, there are:

Lt. Col. R. W. Cook of Muskegon, Mich., Operations Officer for one of the production areas; Lt. Col. W. P. Cornelius of Ennis, Texas, Construction Officer on one of the production plants; Lt. Col. M. C. Fox, of Brooksville, Ohio,

Construction Officer on one of the production plants; Lt. Col. H. L. Friedell of Minneapolis, Minn., Executive Officer of the Medical Section;

Lt. Col. P. L. Guarin, of Houston, Texas, Area Engineer in one of the New York Areas; Lt. Col. J. S. Hodgson, of Montgomery, Ala., Construction Officer on one of the production plants; Lt. Col. A. C. Johnson, of Brooklyn, N. Y., who, as liaison officer in Washington, D. C., was in charge of procurement for the District; Major W. A. Bonnet, Assistant to Colonel Hodgson;

Lt. Col. H. R. Kadlec, now deceased, of Chicago, who was Chief of Construction at the Hanford Engineer Works; Lt. Col. R. W. Lockridge, of Hyattsville, Maryland, technical assistant to the Area Engineer at Santa Fe; Lt. Col. C. A. Nelson, of Pine Bluff, Ark., Director of Personnel for the District;

Lt. Col. W. B. Parsons, of Seattle, Wash., Chief of the Security Division of the District; Lt. Col. A. V. Peterson, of Brooklyn, N. Y., former Area Engineer on the research and development of the Hanford Project and now chief of the Production and Combined Operations Section of the District; Lt. Col. C. E. Rea, of St. Paul, Minn., the head of the Oak Ridge, Tenn., hospital;

Lt. Col. B. T. Rogers, of Eau Claire, Wis., Chief of Construction and Deputy Area Engineer at Hanford Engineer Works; Lt. Col. J. R. Ruhoff, of St. Louis, Mo., who participated in the early scientific developments for the manufacture of basic materials and became Area Engineer for the supply of such materials; Lt. Col. S. L. Stewart of Bisbee, Ariz., Area Engineer for procurement at Santa Fe;

Lt. Col. J. C. Stowers, of Natchez, Miss., Area Engineer on the design of one of the production plants; Lt. Col. A. Tammaro, of Providence, R. I., who was Area Engineer on the manufacture of certain materials for one of the production plants;

Lt. Col. C. Vanden Bulck, of Lincoln Park, N. J., Chief of the Administrative Division of the District; Lt. Col. W. J. Williams, of Spaulding, Ala., who was construction officer in charge of one of the production plants; Major J. O. Ackerman, of Hastings, Minn., Technical Officer on some of the operations at Santa Fe;

Major E. J. Bloch, of St. Louis, who was Unit Chief on coordinating the design, construction and administration of the town of Oak Ridge; Major T. J. Evans, of Florence, Ala., Unit Chief on construction and operation of one of the production plants; Major J. L. Ferry, of Whiting, Ind., head of the industrial division of the medical section; Major H. A. Fidler, of Cambridge, Mass., Area Engineer on the research and development for one of the production processes;

Major O. H. Greager, of Wilmington, Del., who was research division head at the Clinton Laboratories at Oak Ridge with full responsibility for the development of one of the main processes to be used on the Hanford project; Lt. Commander T. M. Keiller, USNR, of Houston, Texas, who was in charge of naval personnel serving in a technical capacity at Clinton Engineer Works; Major W. E. Kelley, of New Albany, Ind., who was unit chief on one of the production plants during design, construction and initial operations;

Major E. J. Murphy, of New York City, who was operations officer of the Clinton Laboratories and related research and experimental work; Major G. W. Russell, of Teaneck, New Jersey, who assisted in the procurement and manufacture of certain basic materials; Major J. F. Sally, of Malverne, Long Island, who was Area Engineer on the design

[158]

and construction of one highly specialized phase of operation;

Major W. T. St. Clair, of Nashville, Tenn., assistant to the construction officer on one of the production plants; Major W. O. Swanson of Jamestown, N. Y., Area Engineer on the design of one of the production plants and also chief of the Utilities and Maintenance Branch at the Clinton Engineer Works; Major H. S. Traynor of Syracuse, N. Y., Production Officer on one technical phase of the project and also chief of the Historical Section; Major J. E. Vance, of New Haven, Conn., Executive and Technical Officer on the procurement of basic materials;

Capt. S. S. Baxter, of Philadelphia, Pa., Area Engineer on medical and scientific research at the University of Rochester; Capt. J. H. King of Anniston, Ala., who was area engineer on one production phase; Capt. A. G. Scheidenhelm, of Mendota, Ill., Commanding Officer of WAC personnel assigned to the District; Capt. B. G. Seitz, of Buffalo, N. Y., statistics officer; Lt. J. J. Flaherty (JG) USNR, of Battle Creek, Mich., deputy chief of the personnel division and Chief Warrant Officer M. S. Levine of Brooklyn, N. Y., who supervised the administration of selective service deferments.

Colonel Nichols said that civilian personnel with the District who made outstanding contributions are:

J. C. Clarke of Philadelphia, Pa., assistant to the chief of the Administrative Division; J. G. LeSieur of Lilbourne, Missouri, chief of the general administrative branch of the Administrative Division; J. R. Maddy, of Enid, Okla., chief of the safety and accident prevention branch of the District; E. A. Wende, of Buffalo, N. Y., chief of the engineering section of the Central Facilities Division at the Clinton Engineer Works and Dr. H. T. Wensel of Washington, D. C., technical advisor to the District Engineer on research under the supervision of the District.

An Accounting of the Successful Operation of the First Self-Sustaining Nuclear Chain Reactor in World History * — Officially Released December 1, 1946

On December 2, 1942, man first initiated a self-sustaining nuclear chain reaction, and controlled it.

Beneath the West Stands of Stagg Field, Chicago, late in the afternoon of that day, a small group of scientists witnessed the advent of a new era in science. History was made in what had been a squash court.

Precisely at 3:25 P.M., Chicago time, scientist George Weil withdrew the cadmium plated control rod and by his action man unleashed and controlled the energy of the atom.

As those who witnessed the experiment became aware of what had happened, smiles spread over their faces and a quiet ripple of applause could be heard. It was a tribute to Enrico Fermi, Nobel Prize winner, to whom, more than to any other person, the success of the experiment was due.

Fermi, born in Rome, Italy, on September 29, 1901, had been working with uranium for many years. In 1934 he bombarded uranium with neutrons and produced what appeared to be both element 93 (uranium is element 92) and also element 94. However, after closer examination it seemed as if nature had gone wild . . . several other elements were present, but none could be fitted into the periodic table near uranium—where Fermi knew they should have fitted if they had been the transuranic elements 93 and 94. It was not until five years later that anyone, Fermi included, realized he

---

* Written for the Manhattan Project by E. R. Trapnell and Corbin Allardice.

had actually caused fission of the uranium and that these unexplained elements belonged back in the middle part of the periodic table.

Fermi was awarded the Nobel Prize in 1938 for his work on transuranic elements. He and his family went to Sweden to receive the prize. The Italian Fascist press severely criticized him for not wearing a Fascist uniform and failing to give the Fascist salute when he received the award. The Fermis never returned to Italy.

From Sweden, having taken most of his personal possessions with him, Fermi proceeded to London and thence to America where he has remained ever since.

An outsider, looking into the squash court where the experiment took place, would have been greeted by a strange sight. Shrouded on all but one side by a grey balloon cloth envelope, was a pile of black bricks and wooden timbers. During the construction of this crude appearing vitally important "pile"—the name that has since been applied to all such devices—the standing joke among those working on it was: "If people could see what we're doing with a million-and-a-half of their dollars, they'd think we were crazy. If they knew why we were doing it, they'd be sure we were."

In relation to the fabulous atomic bomb program, of which the Chicago Pile experiment was a key part, the successful result reported on December 2 formed one more piece for the jigsaw puzzle which is atomic energy. Confirmation of the chain reactor studies was an inspiration to the leaders of the bomb project, and reassuring at the same time, because the Army's Manhattan Engineer District had moved ahead on many fronts. Contract negotiations were under way to build production-scale nuclear chain reactors, land had been acquired at Oak Ridge, Tennessee, and millions of dollars had been obligated.

Three years before the December 2 experiment, it had been discovered that when an atom of uranium was bombarded by neutrons, the uranium atom sometimes split, or fissioned into two parts. Later, it had been found that when an atom of uranium fissioned, additional neutrons were emitted and became available for further reaction with other uranium atoms. These facts implied the possibility of a chain reaction, similar in certain respects to the reaction which is the source of the sun's energy. The facts further indicated that if a sufficient quantity of uranium could be brought together under the proper conditions, a self-sustaining chain reaction would result. This quantity of uranium necessary for a chain reaction under given conditions is known as the critical mass, or, as more commonly referred to, the "critical size" of the particular pile.

For three years the problem of a self-sustaining chain reaction had been assiduously studied. On a cold afternoon nearly a year after Pearl Harbor, a pile of critical size was finally constructed. It worked. A self-sustained nuclear chain reaction was a reality.

Years of scientific effort and study lay behind this demonstration of the first self-sustaining nuclear chain reaction. The story goes back at least to the fall of 1938 when two German scientists, Otto Hahn and Fritz Strassman, working at the Kaiser Wilhelm Institute in Berlin found barium in the residue material from an experiment in which they had bombarded uranium with neutrons from a radium-beryllium source. This discovery caused tremendous excitement in the laboratory because of the difference in atomic mass between the barium and the uranium. Previously, in residue material from similar experiments, elements other than uranium had been found, but they differed from the uranium by only one or two units of mass. The barium differed by approximately

98 units of mass. The question was, where did this element come from? It appeared that the uranium atom when bombarded by a neutron had split into two different elements each of approximately half the mass of the uranium.

Before publishing their work in the German physical journal NATURWISSEN-SCHAFTEN, Hahn and Strassman communicated with Lise Meitner who, having fled the Nazi controlled Reich, was working with Neils Bohr in Copenhagen, Denmark.

Meitner was very much interested in this phenomenon and immediately attempted to analyze mathematically the results of the experiment. She reasoned that the barium and the other residual elements were the result of a fission, or breaking, of the uranium atom. But when she added the atomic masses of the residual elements, she found this total was less than the atomic mass of uranium.

There was but one explanation: The uranium fissioned or split, forming two elements each of approximately half of its original mass, but not exactly half. Some of the mass of the uranium had disappeared. Meitner and her nephew O. R. Frisch suggested that the mass which disappeared was converted into energy. According to the theories advanced in 1905 by Albert Einstein in which the relationship of mass to energy was stated by the equation $E = mc^2$ (energy is equal to mass times the square of the speed of light), this energy release would be of the order of 200,000,000 electron volts - for each atom fissioned.

Einstein himself, nearly 35 years before, had said this theory might be proved by further study of radioactive elements. Bohr was planning a trip to America to discuss other problems with Einstein who had found a haven at Princeton University's Institute of Advanced Studies. Bohr came to America, but the principal item he discussed with Einstein was the report of Meitner and Frisch. Bohr arrived at Princeton on January 16, 1939. He talked to Einstein and J. A. Wheeler who had once been his student. From Princeton the news spread by word of mouth to neighboring physicists, including Enrico Fermi at Columbia. Fermi and his associates immediately began work to find the heavy pulse of ionization which could be expected from the fission and consequent repulse of ionization which could be expected from the fission and consequent release of energy.

Before the experiments could be completed, however, Fermi left Columbia to attend a conference on theoretical physics at George Washington University in Washington, D. C. Here Fermi and Bohr exchanged information and discussed the problem of fission. Fermi mentioned the possibility that neutrons might be emitted in the process. In this conversation, their ideas of the possibility of a chain reaction began to crystallize.

Before the meeting was over, experimental confirmation of Meitner and Frisch's deduction was obtained from four laboratories in the United States (Carnegie Institute of Washington, Columbia, Johns Hopkins, and the University of California). Later it was learned that similar confirmatory experiments had been made by Frisch and Meitner on January 15. Frederic Joliot-Curie in France, too, confirmed the results and published them in the January 30 issue of the French scientific journal, COMPTES RENDUS.

On February 27, 1939, Walter Zinn and Leo Szilard, both working at Columbia University, began their experiments to find the number of neutrons emitted by the fissioning uranium. At the same time, Fermi, and his associates, Herbert L. Anderson and H. B. Hanstein commenced their investigation of the same problem. The results of these experiments were published side-by-side in the April edition of the PHYSICAL REVIEW and showed that a chain reaction might be possible since the uranium emitted additional neutrons when it fissioned.

These measurements of neutron emission by Fermi, Zinn, Szilard, Anderson and Hanstein were highly significant steps toward a chain reaction.

Further impetus to the work on a uranium reactor was given by the discovery of plutonium at the Radiation Laboratory, Berkeley, California, in March 1940. This element, unknown in nature, was formed by Uranium-238 capturing a neutron, and thence undergoing two successive changes in atomic structure with the emission of Beta particles. Plutonium, it was thought might undergo fission if the rare isotope of uranium, U-235 did.

Meanwhile, at Columbia, Fermi and his associates were working to determine operationally possible designs of a uranium chain reactor. Among other things, they had to find a suitable moderating material to slow down the neutrons, since uranium 235 is most readily fissioned by neutrons traveling at relatively low velocities. In July 1941, experiments with uranium were started to obtain measurements of the reproduction factor, (called "K") which was the key to the problem of a chain reaction. If this factor could be made sufficiently greater than 1, a chain reaction could be made to take place in a mass of material of practical dimensions. If it were less than 1, no chain reaction could ocur.

Since impurities in the uranium and in the moderator would capture neutrons and make them unavailable for further reactions and since neutrons would escape from the pile without encountering uranium atoms, it was not known whether a value for "K" greater than unity could ever be obtained.

One of the first things that had to be determined was how

best to place the uranium in the reactor. Fermi and Szilard suggested placing the uranium in a matrix of the moderating material, thus forming a cubical lattice of uranium. This placement appeared to offer the best opportunity for a neutron to encounter a uranium atom. Of all the materials which possessed the proper moderating qualities, graphite was the only one which could be obtained in sufficient quantity of the desired degree of purity.

The study of graphite-uranium lattice reactors was started at Columbia in July 1941, but after reorganization of the entire uranium project in December 1941, Arthur H. Compton was placed in charge of this phase of the work, under the Office of Scientific Research and Development, and the chain reactor program was concentrated at the University of Chicago. Consequently, early in 1942 the Columbia and Princeton groups were transferred to Chicago where the Metallurgical Laboratory was established.

In a general way the experimental nuclear physics group under Fermi was primarily concerned with getting a chain reaction going, the chemistry division organized by F. H. Spedding (later in turn under S. K. Allison, J. Franck, W. C. Johnson, and T. Hogness) with the chemistry of plutonium and with separation methods, and the theoretical group under E. Wigner with designing production piles. However, the problems were intertwined and the various scientific and technical aspects of the fission process were studied in whatever group seemed best equipped for the particular task.

At Chicago, the work on sub-critical size piles was continued. By July 1942 the measurements obtained from these experimental piles had gone far enough to permit a choice of design for a test pile of critical size. At that time, the dies for the pressing of the uranium oxides were designed by Zinn and ordered made. It was a fateful step, since the entire construction of the pile depended upon the shape and size of the uranium pieces.

It was necessary to use uranium oxides because metallic uranium of the desired degree of purity did not exist. Although several manufacturers were attempting to produce the uranium metal, it was not until November that any appreciable amount was available. At that time, Westinghouse Electric and Manufacturing Company, Metal Hydrides Company, and F. H. Spedding, who was working at Iowa State College at Ames, Iowa, delivered several tons of· the highly purified metal and it was placed in the pile, as close to the center as possible. The procurement program for moderating material and uranium oxides had been handled by Norman Hilberry. R. L. Doan headed the procurement program for pure uranium metal.

Although the dies for the pressing of the uranium oxides were designed in July, additional measurements were necessary to obtain information about controlling the reaction, to revise estimates as to the final critical size of the pile, and to develop other data. Thirty experimental sub-critical piles were constructed before the final pile was completed.

Meantime, in Washington, early in 1942 Dr. Vannevar Bush, Director of the Office of Scientific Research and Development, had recommended to President Roosevelt that a special Army Engineer organization be established to take full responsibility for the development of the atomic bomb. During the summer the Manhattan Engineer District was created, and early in September 1942, Major General L. R. Groves assumed command.

Construction of the main pile started in November. The Chicago project gained momentum, with machining of the graphite blocks, pressing of the uranium oxide pellets, and the design of· instruments. Fermi's two construction crews, one under Zinn and the other under Anderson, worked almost around the clock. V. C. Wilson headed up the instrument work.

Original estimates of the critical size of the pile were pessimistic. As a further precaution, it was decided to enclose the pile in a balloon cloth bag which could be evacuated to remove the neutron-capturing air.

This balloon cloth bag was constructed by Goodyear Tire and Rubber Company. Specialists in designing gas-bags for lighter-than-air craft, the company's engineers were a bit puzzled about the aerodynamics of a square balloon. Security regulations forbade informing Goodyear of the purpose of the envelope and so the Army's new square balloon was the butt of much joking.

The bag was hung with one side left open, in the center of the floor a circular layer of graphite bricks was placed. This and each succeeding layer of the pile was braced by a wooden frame. Alternate layers contained the uranium. By this layer-on-layer construction a roughly spherical pile of uranium and graphite was formed.

Facilities for the machining of graphite bricks were installed in the West Stands. Week after week this shop turned out graphite bricks. This work was done under the direction of Zinn's group, by skilled mechanics led by millwright August Knuth. In October, Anderson and his group joined Zinn's men.

Describing this phase of the work, Albert Wattenberg, one of Zinn's group, said: "We found out how coal miners feel. After eight hours of machining graphite, we looked as if we were made up for a minstrel. One shower would only remove the surface graphite dust. About a half-hour after the first shower the dust in the pores of your skin would start oozing.

"Walking around the room where we cut the graphite was

like walking on a dance floor. Graphite is a dry lubricant, you know, and the cement floor covered with graphite dust was slippery."

Before the structure was half complete measurements indicated that the critical size at which the pile would become self-sustaining was somewhat less than had been anticipated in the design.

Day after day the pile grew toward its final shape. And as the size of the pile increased, so did the nervous tension of the men working on it. Logically and scientifically they knew this pile would become self-sustaining. It had to. All the measurements indicated that it would. But still the demonstration had to be made. No matter how well planned, there is always a chance that an experiment will not fulfill expectations. So, as the eagerly awaited moment drew nearer, they gave greater and greater attention to details, the accuracy of measurements, and exactness of their construction work.

Guiding the entire pile construction and design was the nimble-brained Fermi, whose associates describe him as "completely self-confident but wholly without conceit."

So exact were Fermi's calculations, based on the measurements taken from the partially finished pile, that days before its completion and demonstration on December 2, he was able to predict almost to the exact brick the point at which the reactor would become self-sustaining.

But with all their care and confidence few in the group knew the extent of the heavy bets being placed on their success. In Washington, General Groves had proceeded with negotiations with E. I. du Pont de Nemours Company to design, build, and operate a plant based on the principles of the Chicago pile. A "pilot" plant at Oak Ridge and the $350,000,000 Hanford Engineer Works at Pasco, Washington, was to be the result.

At Chicago during the early afternoon of December 1, tests indicated that critical size was rapidly being approached. At 4 P.M., Zinn's group was relieved by the men working under Anderson. Shortly afterwards, the last layer of graphite and uranium bricks was placed on the pile. Zinn, who remained, and Anderson made several measurements of the activity within the pile. They were certain that when the control rods were withdrawn, the pile would become self-sustaining. Both had agreed, however, that should measurements indicate the reaction would become self-sustaining when the rods were withdrawn, they would not start the pile operating until Fermi and the rest of the group could be present. Consequently, the control rods were locked and further work was postponed until the following day.

That night the word was passed to the men who had worked on the pile that the trial run was due the next morning.

About 8:30 on the morning of Wednesday, December 2, the group began to assemble in the squash court.

At the north end of the squash court was a balcony about ten feet above the floor of the court. There the largest part of the observers stayed. Fermi, Zinn, Anderson and Compton were grouped around an instrument console at the east end of the balcony. The remainder of the observers were crowded on the rest of the balcony. R. G. Nobles, one of the young scientists who worked on the pile put it this way: "The control cabinet was surrounded by the 'big wheels'; us 'little wheels' had to stand back."

On the floor of the squash court, just beneath the balcony, stood George Weil, whose duty it was to handle the final control rod. In the pile were three sets of control rods. One set was automatic and could be controlled from the balcony. Another was an emergency safety rod. Attached to one end of this rod was a rope running through the pile, weighted heavily on the opposite end. The rod was withdrawn from the pile and tied by rope to the balcony. Hilberry was ready to cut the rope with an axe should something unexpected happen, or in case the automatic safety rods failed. The third rod, operated by Weil, was the one which actually held the reaction in check until the rod was withdrawn the proper distance.

Since this demonstration was new and different from anything ever done before, complete reliance was not placed on mechanically operated control rods. Therefore, a "liquid-control squad," composed of Harold Lichtenberger, W. Nyer and A. C. Graves, stood on a platform above the pile. They were prepared to flood the pile with cadmium-salt solution in case of mechanical failure of the control rods.

Each group rehearsed what they had to do during the experiment.

At 9:54 Fermi ordered the electrically operated control rods withdrawn. The man at the controls threw the switch to withdraw them. A small motor whined. All eyes watched the lights which indicated the rods' position.

But quickly, the balcony group turned to watch the counters, whose clicking stepped up after the rods were out. The indicators of these counters resembled the face of a clock, with "hands" to indicate neutron count. Near-by was a recorder, whose quivering pen traced the neutron activity with the pile.

Shortly after ten o'clock, Fermi ordered the emergency rod, called "Zip," pulled out and tied.

"Zip out," said Fermi. Zinn withdrew "Zip" by hand and tied it to the balcony rail. Weil stood ready by the "vernier" control rod which was marked to show the number of feet and inches which remained within the pile.

[171]

At 10:37 Fermi, without taking his eyes off the instruments, said quietly:

"Pull it to 13 feet, George." The counters clicked faster. The graph pen moved up. All the instruments were studied, and computations were made.

"This is not it," said Fermi. "The trace will go to this point and level off." He indicated a spot on the graph. In a few minutes the line came to the indicated point and did not go above that point. Seven minutes later Fermi ordered the rod out another foot.

Again the counters stepped up their clicking, the graph pen edged upwards. But the clicking was irregular. Soon it levelled off, as did the thin line of the pen. The pile was not self-sustaining—yet.

At 11 o'clock, the rod came out another six inches; the result was the same: an increase in rate, followed by the levelling-off.

Fifteen minutes later, the rod was further withdrawn and at 11:25 was moved again. Each time the counters speeded up, the pen climbed a few points. Fermi predicted correctly every movement of the indicators. He knew the time was near. He wanted to check everything again. The automatic control rod was reinserted without waiting for its automatic feature to operate. The graph line took a drop, the counters slowed abruptly.

At 11:35, the automatic safety rod was withdrawn and set. The control rod was adjusted and "Zip" was withdrawn. Up went the counters, clicking, clicking, faster and faster. It was the clickety-click of a fast train over the rails. The graph pen started to climb. Tensely, the little group watched, and waited, entranced by the climbing needle.

Whrrump! As if by a thunder clap, the spell was broken. Every man froze—then breathed a sigh of relief when he realized the automatic rod had slammed home. The safety point at which the rod operated automatically had inadvertently been set too low.

"I'm hungry," said Fermi. "Let's go to lunch."

Perhaps, like a great coach, Fermi knew when his men needed a "break."

It was a strange "between halves" respite. They got no pep talk. They talked about everything else but the "game." The redoubtable Fermi, who never says much, had even less to say. But he appeared supremely confident. His "team" was back on the squash court at 2:00 P.M. Twenty minutes later, the automatic rod was reset and Weil stood ready at the control rod.

"All right, George," called Fermi, and Weil moved the rod to a predetermined point. The spectators resumed their watching and waiting, watching the counters spin, watching the graph, waiting for the settling down and computing the rate of rise of reaction from the indicators.

At 2:50 the control rod came out another foot. The counters nearly jammed, the pen headed off the graph paper. But this was not it. Counting ratios and the graph scale had to be changed.

"Move it six inches," said Fermi at 3:20. Again the change —but again the levelling off. Five minutes later, Fermi called:

"Pull it out another foot."

Weil withdrew the rod.

"This is going to do it," Fermi said to Compton, standing at his side. "Now it will become self-sustaining. The trace will climb and continue to climb. It will not level off."

Fermi computed the rate of rise of the neutron counts over a minute period. He silently, grim-faced, ran through some calculations on his slide rule.

In about a minute he again computed the rate of rise. If the rate was constant and remained so, he would know the reaction was self-sustaining. His fingers operated the slide rule with lightning speed. Characteristically, he turned the rule over and jotted down some figures on its ivory back.

Three minutes later he again computed the rate of rise in neutron count. The group on the balcony had by now crowded in to get an eye on the instruments, those behind craning their necks to be sure they would know the very instant history was made. In the background could be heard William Overbeck calling out the neutron count over an annunciator system. Leona Marshall—the only girl present —Anderson, and William Sturm were recording the readings from the instruments. By this time the click of the counters was too fast for the human ear. The clickety-click was now a steady brrrr. Fermi, unmoved, unruffled, continued his computations.

"I couldn't see the instruments," said Weil. "I had to watch Fermi every second, waiting for orders. His face was motionless. His eyes darted from one dial to another. His expression was so calm it was hard. But suddenly, his whole face broke into a broad smile."

Fermi closed his slide rule—

"The reaction is self-sustaining," he announced quietly, happily. "The curve is exponential."

The group tensely watched for twenty-eight minutes while the world's first nuclear chain reactor operated.

The upward movement of the pen was leaving a straight line. There was no change to indicate a levelling off. This was it.

"O.K., 'Zip' in," called Fermi to Zinn who controlled that rod. The time was 3:53 P.M. The rod entered the pile. Abruptly, the counters slowed down, the pen slid down across the paper. It was all over.

Man had initiated a self-sustaining nuclear reaction—**and** then stopped it. He had released the energy of the atom, and controlled it.

Right after Fermi ordered the reaction stopped, the Austrian-born theoretical physicist Eugene Wigner presented him with a bottle of Chianti wine. All through the experiment Wigner had kept this wine hidden behind his back.

Fermi uncorked the wine bottle and sent out for paper cups so all could drink. He poured a little wine in all the cups, and silently, solemnly, without toasts, the scientists raised the cups to their lips—Fermi, Compton, Wigner, Zinn, Szilard, Anderson, Hilberry and a score of others. They drank to success—and to the hope they were the first to succeed.

A small crew was left to straighten up, lock controls, and check all apparatus. As the group filed from the West Stands, one of the guards asked Zinn:

"What's going on, Doctor, something happen in there?"

He didn't hear the report which had gone to General Groves nor the message which Arthur Compton was giving James B. Conant at Harvard, by long distance telephone. Their code was not prearranged.

"The Italian navigator has landed in the New World," said Compton.

"How were the natives?" asked Conant.

"Very friendly."

List of Those Present at "Chicago Pile" Experiment, December 2, 1942

H. M. Agnew, Denver, Colo.
S. K. Allison, Chicago
H. L. Anderson, Chicago
H. M. Barton, Bartlesville, Okla.
T. Brill, Chicago

R. F. Christy, Pasadena, Calif.
A. H. Compton, St. Louis
E. Fermi, Chicago
R. J. Fox, Bentonville, Ark.
S. A. Fox, Bentonville, Ark.
D. K. Froman, Denver, Colo.
A. C. Graves, Los Alamos, N. Mex.
C. H. Greenewalt, Wilmington, Del.
N. Hilberry, Chicago
D. L. Hill, Corinth, Miss.
W. H. Hinch, Denver, Colo.
W. R. Kanne, Schenectady, N. Y.
P. G. Koontz, Fort Collins, Colo.
H. E. Kubitschek, Maywood, Ill.
H. V. Lichtenberger, Chicago
Mrs. L. Woods Marshall, Chicago
G. Miller, Chicago
G. Monk, Jr., New York City
H. W. Newson, Lawrence, Kans.
R. G. Nobles, Willow Springs, Ill.
W. E. Nyer, Chicago
W. P. Overbeck, Richland, Wash.
H. J. Parsons, Chicago
L. Sayvetz, New York City
G. S. Pawlicki, Chicago
L. Seren, Schenectady, N. Y.
L. A. Slotin, Winnipeg, Can. (deceased)
F. H. Spedding, Ames, Iowa
W. J. Sturm, Chicago
L. Szilard, Chicago
A. Wattenberg, New York City
R. J. Watts, Denver, Colo.
G. L. Weil, New York City
E. P. Wigner, Oak Ridge, Tenn.
M. Wilkening, Chicago
V. C. Wilson, Chicago
W. H. Zinn, Chicago

Biographies

ARTHUR HOLLY COMPTON, now chancellor of Washington University at St. Louis, and former dean of the division of physical sciences at the University of Chicago, is probably the world's foremost experimentalist in the field of radiant energy. He received the Nobel Prize in Physics in 1927, making him the third physicist in American history to receive the award. He joined the University of Chicago in 1923, and in 1940 was made dean. For the period of 1941–45 he was in charge of the Metallurgical Project of the Manhattan Project.

ENRICO FERMI, self-exiled Italian physicist, consultant to the Argonne National Laboratory and professor of physics at the University of Chicago, received the Nobel Prize in 1938. He was cited by the War Department as the first man to achieve nuclear chain reaction. During the war, he was associate director of the Los Alamos Laboratory. Fermi was born in Rome and was professor of theoretical physics at the University of Rome from 1927 to 1938, when he left the country because of opposition to Fascism. He was the first to systematize the science of physics in Italy. Mr. Fermi studied at the University of Pisa, Italy, from 1918–22, and has honorary degrees from the Universities of Utrecht and Heidelberg. Before coming to Chicago with the Metallurgical Laboratory, Fermi worked at Columbia University, New York.

WALTER H. ZINN, director of the Argonne National Laboratory, was one of the original members of the Fermi group to work on chain reactors. Zinn was born in Kitchener, Ontario, Canada, in 1906. He received his bachelor's and master's degrees from Queen's University in Canada and his doctor's degree from Columbia University, New York, New York. He taught at Columbia, and City College, New York, before coming to Chicago with the Metallurgical Laboratory.

With Leo Szilard he performed early experiments showing that neutrons are emitted in the fission process; this work became fundamental in studies on atomic energy. Zinn was in charge of a group which constructed the first chain reacting pile and later supervised the design and construction of the first pile using heavy water as the moderator.

HERBERT L. ANDERSON, assistant professor in physics in the Institute for Nuclear Studies, University of Chicago, received his bachelor of science, bachelor of arts and doctor of philosophy degrees from Columbia University. On the atomic bomb project, Anderson did research on nuclear chain reactors with the original Fermi group at Columbia University at Chicago and Los Alamos.

LEO SZILARD, internationally known physicist, who was instrumental in getting President Franklin D. Roosevelt interested in the atomic energy field, is professor of biophysics and professor of social sciences at the University of Chicago. He began his work in the field of nuclear physics in 1934 in London and later continued his work at the University of London. Szilard worked with Enrico Fermi, Nobel Prize physicist, on the early phases of work on chain reaction at Columbia University and at the Metallurgical Laboratory at the University of Chicago. He was born in Budapest, Hungary, in 1898. Szilard received his Ph.D. from the University of Berlin in 1922 and served on the University's faculty there from 1925 to 1933. He became an American citizen in 1943.

NORMAN HILBERRY, associate director of the Argonne National Laboratory, was one of the scientists who worked on the December 2 pile. His was the responsibility for procuring moderator material and uranium oxide for the reactor. Born in Cleveland, Ohio, in 1899, Hilberry received his bachelor's degree from Oberlin College, Ohio, and his Ph.D. in physics from the University of Chicago. He taught

[178]

at the University of Chicago, and New York University. He is a fellow of the the New York Academy of Spectroscopy, and has carried on extensive studies of the discharge of electricity through gases; physical optics; cosmic ray showers; and the constitution of primary cosmic rays and their secondary radiations.

Oak Ridge Organizations

African Violet Club
Alcoholics Anonymous
Altrusa Club
American Cancer Society
American Ceramic Society
American Legion
American Legion Auxiliary
American Physical Society
American Institute of Electrical Engineers
American Red Cross
American Society of Chemical Engineers
American Chemical Society
American Society of Civil Engineers
American Society for Metals
American Society for Mechanical Engineers
American Society of Safety Engineers
Am Vets
Association for Childhood Education
Association of Life Underwriters

Association of Engineers and Scientists
AEC Employee's Association
Air Reserve Association
Atomic Trades and Labor Council (AFL)
Bar Association
Berea College Alumni Club
Beta Sigma Phi Sorority
Boy Scouts of America
Business and Professional Women's Club
Chamber of Commerce
Camera Club
Cat Fancier's Club
Cedar Hill Parents Association
Central Labor Union (AFL)
Chess Club
Children's Theatre
Children's Museum
Cinema Club
Citizens Fund Raising Screening Committee
Civic Music Association
Civil Air Patrol
Community Chest, Inc.

Community Chorus
Community Playhouse
Community Singing
Conversation Club
Coon Hunters
Dance Club
Daughters of American Revolution
De Molay
Disabled American Vets
Duplicate Bridge Club
Eagles
Eastern Star
Elks
Eye of Americans
Exchange Club
Fishing Club
Fraternal Order of Police
Georgia Tech Alumni Association
Girl Scouts Council
Gray Lady Corps
Hiking Club
High School Service Club
Holy Name Society
Instrument Society of America
Jaycettes
Junior Chamber of Commerce
Junior Red Cross
Kennel Club
Kiwanis International
Knights of Columbus
League of Women Voters
Library Board
Lions Club
Lioness Club

Masons
Medical Society
Medical Society Auxiliary
Men of the United Church
Michigan Alumni Club
Ministerial Association
Model Airplane Club
Model Railroad Club
Moose
Municipal Band
Music Listening Group
National Council of Jewish Women
National Federation of Federal Employees
National Guard
Nurses Association
Oak Ridge Golf and Country Club
Oak Ridge Education Association
Oak Ridge Sportsman Association
Oak Ridge Yacht Club
Oak Ridge Advisory Planning Commission
Parent Teachers Association
P. E. O. Sisterhood
Pistol Club
Photographers Association
Power Engineers
Power Squadron
Public Health Advisory Council
Radio Operators Club
Rainbow Girls

Recreation Board
Reserve Officers Association
Riding Club and Horse Show Association
Rifle Club
Rotary Club
Ski Club
Shrine Club
Stamp Club
Symphony Orchestra
Tennis Club
Town Council
Trap and Skeet Club
Trade and Industrial Club
Tuberculosis Association
United Gas, Coke and Chemical Workers (CIO)
United World Federalists
Veterans of Foreign Wars of the United States
United Veterans Association
Veterans of Foreign Wars
Virginia Polytechnic Institute Alumni Association
Virginia Alumni Association
Welfare Council
Welfare Services Advisory Council
Wildcats Den
Women's Bowling Association
Women's Club
Women's Democrat Club
Young Democrats Club
Young Republican Club

Churches

Baptist, Calvary
Baptist, First
Baptist, Freewill
Baptist, Glenwood
Baptist, Highland View
Baptist, Mt. Zion
Baptist, Robertsville
Catholic, Saint Mary's
Christian Science
Church of Christ, Cedar Hill
Church of Christ, Highland View
Church of Jesus Christ of the Latter-Day Saints
Church of the Open Door
Church of God, Iroquois
Community Church
Episcopal, St. Stephens
First Christian Church
Jewish
Lutheran, Faith
Lutheran (Grace Evangelical)
Methodist, First
Methodist, Kern Memorial
Methodist, Trinity
Methodist Episcopal Church
Nazarene, Church of The
Presbyterian, First
United Church
United Church, Oak Valley
Unitarian, Tennessee Valley

Ingram Content Group UK Ltd.
Milton Keynes UK
UKHW022228060723
424626UK00005B/56